Progress in IS

W0080335

More information about this series at http://www.springer.com/series/10440

Nguyen Hoang Thuan

Business Process Crowdsourcing

Concept, Ontology and Decision Support

 Springer

Nguyen Hoang Thuan
Faculty of Information Technology
Can Tho University of Technology
Can Tho City
Vietnam

This book is based on a doctoral thesis successfully defended at the Victoria University of Wellington.

ISSN 2196-8705   ISSN 2196-8713   (electronic)
Progress in IS
ISBN 978-3-030-08241-3   ISBN 978-3-319-91391-9   (eBook)
https://doi.org/10.1007/978-3-319-91391-9

Printed on acid-free paper

This Springer imprint is published by the registered company Springer International Publishing AG part of Springer Nature
The registered company address is: Gewerbestrasse 11, 6330 Cham, Switzerland

Foreword

The idea of using technology to develop collective intelligence has for long been explored by science fiction. A wonderful example is the Lensman series written by the great science fiction pioneer E. E. "doc" Smith between 1934 and 1954 (Smith 1948). Lensman were committed the almost impossible mission to save the universe from being subjugated by the Eddorians. To help them realize that mission, the Arisians gave them a tool: The Lens. Lenses expand the mental skills of their wearers with capabilities such as communicating across species, mind reading, telepathy and augmented thinking. Many movies have also explored this idea. For instance, the Jedi in Star Wars have the Force, which binds them together through a common, persistent conscience and a communication medium. Avatar showed us Eywa, a biosphere that supports a planet-scale network of living entities, which functions as our brain's neural network on a grander scale (Baxter 2012).

More down to earth but not less exciting, the pursuit of collective intelligence has also taken a great place in science and technology. Vannevar Bush, in the influential "As We May Think" essay published in 1945, proposed the Memex, a machine capable to expand the human mind by preserving personal records and communications (Bush 1945). At the time, it was impossible to implement the Memex (which required using miniature cameras and microfilm), but it inspired others like Doug Engelbart and Ted Nelson. Doug Engelbart developed several first-of-a-kind pieces of technology like the first computer mouse, the first working hypertext system and the first collaborative system (Engelbart and English 1968). Ted Nelson's Xanadu was so conceptually advanced that it has not yet materialized (Nelson 1982). Xanadu was supposed to manage an information Web using bidirectional links, which provide more powerful searches than we can do today using Web browsers (Knowlton 2015). Nevertheless, these ideas inspired the development of the Internet, World Wide Web and many other admirable projects like Wikipedia and GitHub, which bring together and promote our collective intelligence (Berners-Lee et al. 2010; Smith and Weiss 1988).

Technological advances in human–computer interaction have also encouraged the pursuit of collective intelligence. Worth mentioning is the idea that technology has value beyond mechanization and automation. The computer not only does things for us. It can be an amplification of us, as suggested by concepts such as embodied interaction (Dourish 2001), joint cognition (Hollnagel and Woods 2005) and bricolage (Cabitza and Simone 2015).

Our organizations have been evolving to explore and exploit technological innovations. Electricity, fridges and elevators contributed immensely to aggregate people in specialized, complex and interdependent urban structures. Then cars, phones, computers and many other technologies contributed further to intensification, automation, decentralization and collaboration (Stott 1992). Nowadays, organizations are absorbing the impacts of constant interconnectivity, powerful mobile technology and embedded systems. Organizations are becoming more virtual, ubiquitous, agile, information rich and of course more complex (Alberts 2011).

Crowdsourcing emerges as another conceptually simple but disruptive technology capable to change significantly the structure and behaviour of our organizations through collective intelligence. Initially, it may have been regarded as another way to execute one-off projects, to solve simple problems using brute force, or maybe another way to outsource certain functions at low cost. But that is just the beginning of the story. The fully integrated, continuous and dynamic use of crowdsourcing may turn organizations less structural, bounded and predictable. Furthermore, crowdsourcing creates the opportunity to jump-start new activities, to bring in a continuous flow of ideas, expertise and knowledge, and to change strategic directions without much attrition. And all independently of time, space and size, crowdsourcing may confer elasticity, malleability, agility, scale, and resilience.

Right now, we are just seeing the initial steps towards integrating crowdsourcing into organizations. Much research and development are still necessary to fully understand the value and the potential uses and consequences. This book by Nguyen Hoang Thuan is a great step in that direction. It brings together a large body of knowledge on the subject, adding to it an integrated business perspective, which provides a solid foundation for understanding where crowdsourcing is today, how it can move forward within our organizations, and how it contributes to that great idea of collective intelligence. We do not yet have the Lens, and neither Eywa, but maybe one day someone may be able to say that crowdsourcing gave a small push in that direction.

Wellington, New Zealand   Assoc. Prof. Pedro Antunes

References

Alberts, D. (2011). *The agility advantage*. CCRP.
Baxter, S. (2012). *The science of Avatar*. Orbit.
Berners-Lee, T., Cailliau, R., Groff, J., & Pollermann, B. (2010). World-wide web: The information universe. *Internet Research, 20*(4), 461–471.
Bush, V. (1945). As we may think. *The Atlantic Monthly*.

Cabitza, F., & Simone, C. (2015). Building socially embedded technologies: Implications about design. In *Designing socially embedded technologies in the real-world* (pp. 217–270). London: Springer.

Dourish, P. (2001). *Where the action is*. Cambridge, MS: The MIT Press.

Engelbart, D., & English, W. (1968). A research center for augmenting human intellect. In *Proceedings of the December 9–11, 1968, Fall Joint Computer Conference, Part I* (pp. 395–410). ACM.

Hollnagel, E., & Woods, D. (2005). *Joint cognitive systems: Foundations of cognitive systems engineering*. Boca Raton, FL: CRC Press.

Knowlton, K. (2015). Ted Nelson's Xanadu. In *Intertwingled* (pp. 25–28). Springer.

Nelson, T. (1982). *Literary machines*. Mindful Press.

Smith, E. (1948). *Triplanetary*. Fantasy Press.

Smith, J., & Weiss, S. (1988). Hypertext. *Communications of the ACM, 31*(7), 816–819.

Stott, R. (1992). Urban electrification. *Reviews in American History, 20*(2), 211–215.

Acknowledgements

My first acknowledgement is to Assoc. Prof. Pedro Antunes and Dr. David Johnstone. You not only showed me how to structure, sharpen and publish research results, but also moved forward my thinking on business process crowdsourcing. Your invaluable support, encouragement, ideas and collaboration contribute to make it possible for me to complete this book.

I would like to thank my family for their support and understanding. Thanks to my parents and my mother-in-law. Thanks to my wife, Xuan Trang, for love, support, and motivation. Thanks, too, to my son, Hoang Bach, providing me with the strength and fun to complete this book.

A special thanks to the University of Wellington (VUW) academics, staff and administrative team. You provided all necessary support for my research in the field of crowdsourcing, and this resulting book. My sincere thanks to the New Zealand ASEAN Scholarship. I am honoured to be one of the scholarship holders. Thanks to the Can Tho University of Technology; you were there to assist me during my data collection. This book was finalized while the author has worked at the Can Tho University of Technology.

Finally, I am grateful to all individuals and organizations that were involved in the research presented in this book.

Contents

Abbreviations

AMT	Amazon mechanical turk
API	Application programming interface
BPC	Business process crowdsourcing
BPM	Business process management
BPMN	Business process model and notation
BPO	Business process outsourcing
CT	Crowd tagging
CTUT	Can Tho University of Technology
DSS	Decision support system
DTN	Design theory nexus
GUI	Graphical user interface
IM	Industrial management
IP	Intellectual property
IS	Information systems
IT	Information technology
LDC	Logo design contest
R&D	Research and development
SDLC	System development life cycle
VUW	Victoria University of Wellington

List of Figures

List of Tables

Chapter 1
Introduction

Welcome to the age of the crowd.
—Jeff Howe

To open this book, let us imagine how excellent organisations could be if they were not limited to human resource constraints and could access labour and skills on demand. In fact, this image has now come true thanks to crowdsourcing. *Crowdsourcing* is an emerging sourcing strategy that utilises Internet users through an open call to perform tasks (Howe, 2006b). This strategy has been used by many organisations for harnessing on-demand workforce, external expertise, knowledge, and creativity. The development of crowdsourcing has been gathering momentum in terms of growth revenues and adoption. For instance, in 2011 crowdsourcing revenue reached \$375.70 million, while increasing 74.7% within one year[1]; between 2013 and 2014 the adoption of crowdsourcing by leading organisations like Microsoft and Google increased 48%[2]; and in 2017 the list of organisations adopting crowdsourcing has become longer, including Procter & Gamble, Unilever, Nestlé, Johnson & Johnson, General Mills, and PepsiCo.[3]

1.1 Research Context

The fundamental idea behind the crowdsourcing strategy is that an organisation (which could be a company, non-profit organisation, or government) defines tasks and broadcasts them online to the crowd, who voluntarily undertake these tasks in an individual or collaborative way. When completing these tasks, individuals in the crowd submit their work back to the organisation, which assesses the work quality and may provide

---

[1]Source: Massolution White Paper: http://www.lionbridge.com/files/2012/11/Lionbridge-White-Paper_The-Crowd-in-the-Cloud-final.pdf.
[2]Source: eYeka Trend Report: http://eyeka.pr.co/99215-eyeka-releases-the-state-of-crowdsourcing-in-2015-trend-report.
[3]Source: eYeka The State of Crowdsourcing in 2017: The Age of Ideation: https://en.eyeka.com/resources/reports?download=cs_report_2017.pdf.

© Springer International Publishing AG, part of Springer Nature 2019
N. H. Thuan, *Business Process Crowdsourcing*, Progress in IS,
https://doi.org/10.1007/978-3-319-91391-9_1

incentives or individual compensations (Ghezzi, Gabelloni, Martini, & Natalicchio, 2017; Zhao & Zhu, 2014). It is worth mentioning that the process normally unfolds through crowdsourcing platforms like InnoCentive[4] and Amazon Mechanical Turk (AMT).[5]

Among the organisations adopting crowdsourcing, the most interesting, and maybe biggest one, was the New Zealand government. In 2015, the government established a $25.7 million project, named the Flag Consideration Project,[6] which looked for the future flag of New Zealand. Interestingly, this project relied neither solely on design companies nor on professional designers, rather it had been opened to everybody. More precisely, the project proposed an open call for all New Zealanders to create flag designs and ultimately they would decide on the future flag of New Zealand. The project started in May 2015 and received more than 10,000 flag designs after three months. These designs were then shortlisted into four alternatives. The shortlisted designs and the current flag were voted by New Zealanders through a two-round referendum in December 2015 and March 2016. More than two million New Zealanders voted in the referendum with the final decision to retain the current flag. For the project, crowdsourcing allowed the New Zealand government to harness creativity and design expertise, to know what New Zealanders stand for, and to collectively make the final decision. In this manner, the crowdsourcing project has already been very successful, receiving a huge number of submissions and attracting significant public attention, as shown in Fig. 1.1.

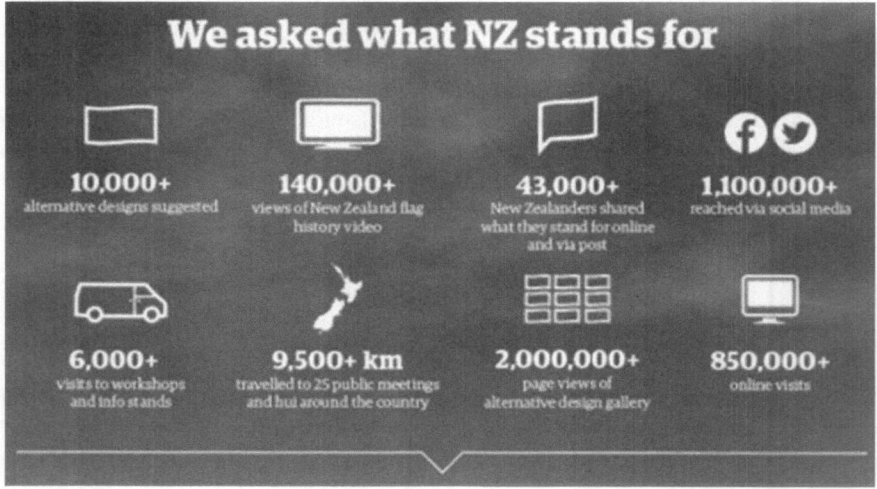

Fig. 1.1 The flag consideration project statistics. *Source* The flag consideration project: http://ndhadeliver.natlib.govt.nz/ArcAggregator/arcView/frameView/IE25848623/https:/www.govt.nz/browse/engaging-with-government/the-nz-flag-your-chance-to-decide/

---

[4]Source: InnoCentive: http://www.innocentive.com/.

[5]Source: Amazon Mechanical Turk: https://www.mturk.com/mturk/welcome.

[6]Source: The Flag Consideration Project: https://www.govt.nz/browse/engaging-with-government/the-nz-flag-your-chance-to-decide/.

Though crowdsourcing has been used to harness creative design skills, as in the Flag Consideration Project, the use of crowdsourcing is not limited to design or to attracting public attention. Organisations have been adopting the crowdsourcing strategy for varied purposes. For instance, BMW, Boeing, Colgate-Palmolive, Procter and Gamble, and Netflix have all used crowdsourcing for research and development (R&D) activities. These organisations have published difficult R&D issues online and have called for innovative solutions from the crowd. Also utilising the crowd, Lego, Threadless, and Starbucks have used crowdsourcing for gathering customers' ideas, which generates innovation and increases the ties with their customers. Other organisations rely on the crowd for performing their day-to-day activities, such as transcribing texts, gathering customer surveys, and processing information. All these examples, and many others,[7] illustrate how organisations have successfully utilised the *crowd* for *sourcing* organisational tasks.

The power of crowdsourcing lies in its ability to distribute work beyond the organisational boundaries and harness a variety of business endeavours. By adopting a crowdsourcing strategy, organisational tasks can be accomplished by Internet users, independent from time and geographic constraints. Crowdsourcing can also harness expertise, problem solving, knowledge and creativity from outside of the organisational boundaries, and serve as a project stimulator, as shown in the Flag Consideration Project. Furthermore, it is a relatively cost-effective way of doing work because the crowd can be assembled on demand and in many cases are voluntary, or receive only a few dollars per task (Brabham, 2013; Saxton, Oh, & Kishore, 2013; Stol, LaToza, & Bird, 2017). All in all, virtually limitless workforce, varied skills and relatively low costs make crowdsourcing a valuable sourcing strategy for organisations.

Bringing value to organisations, crowdsourcing at the same time changes the way organisations perform and manage work. With crowdsourcing, organisations have been able to access human resources from virtually everywhere. This openness requires organisations to re-define their organisational boundaries, and more importantly to effectively establish conduits between internal and external activities across the boundaries (Thuan, Antunes, & Johnstone, 2017; Tranquillini, Daniel, Kucherbaev, & Casati, 2015). Furthermore, crowdsourcing has changed organisational work structures. Using members of the crowds to finish work, crowdsourcing promotes a scalable bottom-up structure, which is different with the stable top-down hierarchy of traditional structures (Brabham, 2013; Kohler, 2015). These changes accordingly reflect an organisational shift from a closed business model to a new crowdsourcing business model.

However, in this new crowdsourcing business model, there is a missing component from the above successful stories. It is clear about what organisations can possibly achieve through crowdsourcing, but how they adapt their traditional business structure to establish a crowdsourcing business model is not yet clear.

---

[7]Source: List of Crowdsourcing Projects: https://en.wikipedia.org/wiki/List_of_crowdsourcing_projects.

Motivated by this 'how to' challenge, coupled with the growing popularity of crowdsourcing, this book aims at helping organisations to successfully establish the crowdsourcing strategy.

1.2   Research Problem and Objectives

The success of the crowdsourcing strategy depends on how organisations establish crowdsourcing processes. A crowdsourcing process is a set of activities that need to be performed to operationalise the strategy (Thuan et al., 2017). Experience shows that well-structured processes are assumed not only to produce better crowdsourcing results, but also to deploy the crowdsourcing strategy in a more manageable way (Tranquillini et al., 2015). In contrast, an ad hoc process prevents crowdsourcing to maximise its benefits due to the need for re-planning and failed outputs (Rouse, 2010). Furthermore, understanding the crowdsourcing process is important to define its information flows, which is an antecedent for crowdsourcing implementation. All in all, the crowdsourcing process plays a central role in the crowdsourcing strategy.

Given the important role, crowdsourcing process has attracted much attention from researchers. This book classifies the research on crowdsourcing processes from two different views: low and high levels of granularity. The low granularity has been adopted by a large number of studies that investigated several aspects of the crowdsourcing process. While offering an important understanding on crowdsourcing processes, most of these studies focus on individual parts of the crowdsourcing process. The ad hoc nature, which makes difficult to understand and establish crowdsourcing processes, has been repeatedly complained in the domain (Amrollahi, 2015; Thuan et al., 2017; Zhao & Zhu, 2014). That is, the domain has few contributions towards a holistic view of the crowdsourcing process. Rather, many of them regard crowdsourcing as a one-off process. As a result, this group of studies have left us with unstructured, scattered, and sometimes conflicting knowledge, which hinders our ability to build a dedicated repeatable crowdsourcing process.

Recently, some research efforts have adopted a high level of granularity when analysing and conceptualising the crowdsourcing process. Seeking an overall view of the crowdsourcing process (Grace et al., 2015; Kucherbaev, Daniel, Tranquillini, & Marchese, 2016; Muhdi, Daiber, Friesike, & Boutellier, 2011), they deal with rather high-level conceptualisation, and thus face significant gaps regarding explanation of how to effectively establish the crowdsourcing process in detail. Furthermore, as they are mainly exploratory efforts, further empirical research is needed to test their propositions before their practical usage. Consequently, the domain still lacks a solid knowledge base that organisations can rely upon to establish the crowdsourcing process.

Together, these challenges indicate that the crowdsourcing process is immature, with a lack of a solid knowledge base, unstructured sets of knowledge sources, and a dominant one-off perspective. This immaturity has prevented organisations from establishing crowdsourcing as a repeatable organisational process. Given that, the central research problem addressed in this book is as follows:

> **Research problem**: The immature approach to crowdsourcing, characterised by a lack of a solid knowledge base and unstructured sets of knowledge sources, has prevented organisations from establishing repeatable crowdsourcing processes.

Addressing this problem enables crowdsourcing to evolve from an unstructured, immature form towards a more structured repeatable process. To this end, it seems that reconciling the low and high levels of granularity provides a more integrated picture of the crowdsourcing process. That is, we can understand both the details of crowdsourcing processes and their abstract coordination. This integrated view suggests the use of a *business process* lens, which has scarcely been adopted in the crowdsourcing field.

This book will use a business process lens to investigate crowdsourcing. Essentially viewing complex processes as a set of independent, yet coordinated, activities (van der Aalst & Hee, 2004), the business process lens allows us to analyse independent crowdsourcing elements and link them into an integrated crowdsourcing process. We designate this particular lens as *Business Process Crowdsourcing* (BPC), the term was first used by La Vecchia and Cisternino (2010). Adopting the BPC view, we seek to consider crowdsourcing as a repeatable business process, overcoming the one-off viewpoint. Further, BPC that relies on both high and low levels of granularity enables us to analyse and structure existing knowledge sources, and to build a solid knowledge base. Consequently, we expect BPC to move crowdsourcing towards a more mature form.

Focusing on the concept of BPC, the book sets four research objectives:

- RO1: The first research objective is to *understand the main building blocks of BPC that can be identified in the domain*, which allows us to conceive the concept of BPC. We define the term 'building blocks', aligning to Osterwalder (2004), as common decomposed elements that can be combined to describe the overarching concept.
- RO2: The next research objective seeks *to develop a model structuring the identified building blocks for conceptualising BPC*. To do this, we structure the identified building blocks into a process model that considers crowdsourcing as a repeatable business process.
- RO3: We aim to construct a solid knowledge base of BPC. The literature has suggested that domain ontologies can consolidate knowledge and construct knowledge bases (Kohlborn, 2012; Miah, 2008). In this vein, we seek *to*

construct a domain ontology of BPC that organises the unstructured knowledge in the domain.

- RO4: If we successfully construct a solid knowledge base of BPC, computer-based decision support can be further developed for assisting organisations establishing their business processes based on crowdsourcing. This leads to the final research objective of the book, which aims *to construct a decision tool supporting organisations in establishing BPC.*

These four research objectives guide this study. To achieve these objectives, we need an appropriate research approach.

1.3 Research Approach

This study adopts a design science paradigm (Hevner, March, Park, & Ram, 2004). Gradually being embraced as an important and solid research paradigm in the Information Systems (IS) discipline, design science focuses on constructing innovative solutions and shows how to do that effectively (Gregor & Hevner, 2013). Consequently, it is a good fit with the 'construct' and 'support' focus of the study. This adoption is further appropriate as design science emphasises a rigorous approach to advance current knowledge on design problems. The knowledge advancement is extremely necessary with unstructured domains like crowdsourcing.

A key principle of design science research is that it must be founded on a rigorous knowledge foundation, which according to Hevner and Chatterjee (2010) comprises three types: (1) scientific theories; (2) meta-artefacts; and (3) experience and expertise. With the ad hoc nature of the domain, we could not find a prevailing crowdsourcing theory or meta-artefact on which to base the research. Thus, the current study focuses on the third type of knowledge foundations set by Hevner and Chatterjee (2010): building a knowledge base from individual knowledge sources predominant in the domain.

Another key principle of design science is that research has to follow an appropriate design method guiding the research activities. For this study, we follow a design method comprising four research stages. First, this method systematically analyses and scopes individual knowledge sources in the domain for understanding the main building blocks of BPC (Paré, Trudel, Jaana, & Kitsiou, 2015). Second, the method proposes a conceptual model using the identified building blocks to conceptualise the phenomenon (Webster & Watson, 2002). Then, it constructs an ontology to consolidate the domain knowledge (Corcho, López, & Gómez-Pérez, 2003). Finally, the ontological view is used to construct a decision tool supporting the establishment of BPC (Lim, Stolterman, & Tenenberg, 2008).

Aligning the tenets of design science, each research stage includes two equally important activities: build and evaluate (Hevner et al., 2004; March & Smith, 1995). Apart from the first stage, scoping and synthesising knowledge sources in order to feed other research activities and thus does not need a separate evaluation, artefacts from the other three stages are carefully evaluated. That is, the conceptual model is assessed through a case study approach (Yin, 2013a). The ontology is evaluated by triangulation. The decision tool is evaluated by experiments (Montgomery, 2012) and focus groups (Tremblay, Hevner, & Berndt, 2010). Together, the iterations of build and evaluate activities constitute the research process of the study. Figure 1.2 presents the research process, consisting of the four research stages and their alignment with the research objectives.

In summary, the research design adopts the guidance for design science research defined by Hevner et al. (2004) and comprises four main stages: scoping knowledge sources, conceptual model, ontology, and decision tool. Each stage includes both the build and evaluate activities, which together constitute the iterative research process. An exception is the first stage, which scopes the knowledge sources for feeding other activities and thus does not need a separate evaluation. These stages are based on rigorous research techniques and relevant data collected from the practical environments.

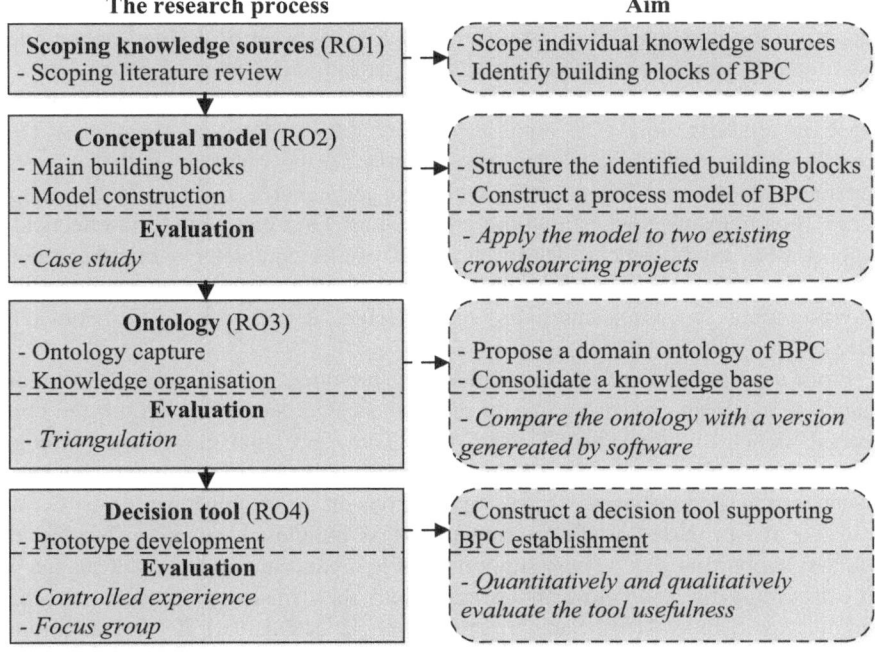

Fig. 1.2 The research process

1.4   Research Significance

Positioned within the crowdsourcing field and design science paradigm, the significance of this study concerns both academics and practitioners. From an academic point of view, this research brings the business process lens to crowdsourcing research, which possibly evolves crowdsourcing from one-off processes to repeatable organisational business processes. Just as the movement of business process outsourcing (PBO) contributed to advance the outsourcing field, BPC is a necessary development for moving the crowdsourcing field forward. Further, this development addresses the ad hoc nature of crowdsourcing processes (Thuan et al., 2017; Zhao & Zhu, 2014) and responds to the calls for developing an integrated crowdsourcing process (Djelassi & Decoopman, 2013; Khazankin, Satzger, & Dustdar, 2012a; Lüttgens, Pollok, Antons, & Piller, 2014).

In the design science paradigm, the contribution of this research is a set of artefacts establishing BPC: a conceptual model, an ontology, and a decision tool. In particular, the model conceptualises BPC and its building blocks. The ontology consolidates the domain knowledge through defining the concepts, hierarchical relationships and decision-making relationships. Together, the model and ontology fulfils the gap of "little attention to the ontological and conceptual foundations on how to engineer the entire [crowdsourcing] process" (Hosseini, Phalp, Taylor, & Ali, 2014, p. 1). The decision tool, operationalising the ontology, supports making informed BPC decisions. Therefore, it expands early efforts in developing decision support in the domain. Overall, the study contributes a set of design science artefacts for establishing crowdsourcing as an organisational business process.

Another contribution of the study is the empirical evidence that shows how the proposed artefacts work. The empirical results, derived from case studies of two crowdsourcing projects, a triangulation comparing the ontology with a version generated by software, experiments with 190 participants, and two focus groups with ten participants, suggest that the proposed artefacts can be used to effectively support BPC establishment. These empirical results complement our theoretical efforts conceptualising BPC, and other theoretical efforts trying to consolidate the crowdsourcing process (Amrollahi, 2015; Geiger & Schader, 2014; Hetmank, 2013).

From a practical point of view, this study provides several means for organisations to establish crowdsourcing in their business processes, including the conceptual model, ontology, and decision tool. The conceptual model and ontology guide how to plan, analyse, and design BPC, while the tool supports business managers and process designers making decisions on the establishment of BPC. We note that the proposed artefacts can be used as individual means or as a set of artefacts supporting BPC. These supports enable organisations to take advantage of crowdsourcing by integrating the strategy into their business processes (Lopez, Vukovic, & Laredo, 2010; Satzger, Psaier, Schall, & Dustdar, 2012; Tranquillini et al., 2015).

1.5 Structure of the Book

This chapter has introduced the concept of business process crowdsourcing, and defined the research problem and research objectives. It also presented the research approach and possible contributions. The remaining chapters of this book are structured as follows. Chapter 2 provides the background of existing research in crowdsourcing. It identifies the main concepts in the crowdsourcing field and presents an overview of related work in important areas of crowdsourcing processes. In Chap. 3, we use a scoping literature review to extract and scope knowledge sources in the domain, which identifies the main building blocks of business process crowdsourcing. Chapter 4 addresses the construction of a process model for establishing business process crowdsourcing. The new process model is empirically evaluated using two case studies. In Chap. 5, we build an ontology that captures main concepts and relationships in the domain. We evaluate the ontology by comparing it with an automated ontology generated by software. Chapter 6, based on the foundation provided by the ontology, builds a decision tool supporting business process crowdsourcing. The tool is evaluated by both controlled experiment and focus group. Finally in Chap. 7, the research results, contributions, implications, limitations, and main conclusions are presented. Future research opportunities are outlined.

This book covers the author's research and the research collaboration between the author and the co-authors over the last five years. The content is fundamentally based on the following journal articles and conference papers. Consequently, the personal pronoun 'we' has been used in the book to reflect the collaboration and in accordance with standard scientific protocol.

- Thuan, N. H., Antunes, P., & Johnstone, D. (2018). A decision tool for business process crowdsourcing: Ontology, design, and evaluation. *Group Decision and Negotiation, 27*(2), 285–312 (Re-use under Springer license - License number: 4301100207257).
- Thuan, N. H., Antunes, P., & Johnstone, D. (2017). A process model for establishing business process crowdsourcing. *Australasian Journal of Information Systems, 21,* 1–21 (Re-use with permission from Thuan et al. (2017)).
- Thuan, N. H., Antunes, P., & Johnstone, D. (2016). Factors influencing the decision to crowdsource: A systematic literature review. *Information Systems Frontiers, 18*(1), 47–68 (Re-use under Springer license - License number: 4245350121048).
- Thuan, N. H., Antunes, P., Johnstone, D., & Ha, X. S. (2015). Building an enterprise ontology of business process crowdsourcing: A design science approach. *The 19th Pacific Asia Conference on Information Systems (PACIS 2015 Proceedings). AISeL, Paper 112* (Re-use with permission from Thuan et al. (2015)).

- Thuan, N. H., Antunes, P., & Johnstone, D. (2014). Toward a nexus model supporting the establishment of business process crowdsourcing. In T. K. Dang, R. Wagner, E. Neuhold, M. Takizawa, J. Küng, & N. Thoai (Eds.), *The 1st International Conference on Future Data and Security Engineering (FDSE 2014). LNCS* (Vol. 8860, pp. 136–150): Springer, Heidelberg (Re-use under Springer license—License number: 4245360639525).

Chapter 2
Background

> *Crowdsourcing research is a dynamic and vibrant research*
> *area, and has been steadily growing over the years.*
> —Zhao and Zhu (2014)

As crowdsourcing has raised multiple interests, it has been studied in a variety of domains: marketing, management, software engineering, computer science, and information systems. This wide research spectrum enables crowdsourcing to become a young yet rapidly growing field. Publications in this field cover aspects like decision making, quality control, crowd management, workflow design, system architecture and crowd programming (Afuah, Tucci, & Viscusi, 2018; Kohler & Nickel, 2017; Zhao & Zhu, 2014). To help readers understand some key aspects of crowdsourcing, this chapter presents a focused literature review of crowdsourcing research.

The variety of crowdsourcing literature makes the body of knowledge hard to be synthesised. To help achieve a shared structure and understanding of the concept, we propose a layered framework that provides separation of concerns. Figure 2.1 presents the framework comprising of four layers: conceptualisation, classification, process and establishment. These layers are structured symmetrically (top to bottom) from being more abstract to more concrete, and from overview to focus on the research phenomenon.

The first layer conceptualises what crowdsourcing is by characterising three major research streams: crowdsourcing underpinnings, related concepts and existing definitions of crowdsourcing. The literature in each stream is reviewed in Sect. 2.1. The second layer examines the classifications of crowdsourcing and its related elements, which are presented in Sect. 2.2. Classifications and taxonomies are focused because they can provide a structured way to organise knowledge in the field (Nickerson, Varshney, & Muntermann, 2012). Among the different elements classified in the literature, the review highlights the applications, tasks, crowd members and platforms as the most pertinent to this book.

The third and four layers are presented in Sect. 2.3 in order to analyse the current state of business process crowdsourcing. It begins with a review of studies on crowdsourcing processes. The two predominant views, low and high levels of

N. H. Thuan, *Business Process Crowdsourcing*, Progress in IS,
https://doi.org/10.1007/978-3-319-91391-9_2

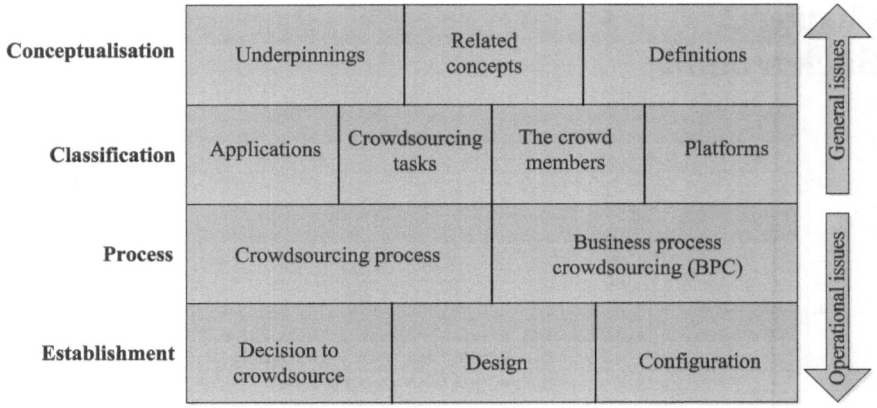

Fig. 2.1 Layered framework for the literature review

granularity for researching crowdsourcing processes, are reviewed. The focus then moves to the more specific concept of BPC. The relatively small body of research related to the concept is reviewed for identifying the important roles of BPC and the need for further investigating BPC. Next, the review analyses the three main stages necessary to establish BPC: decision to crowdsource or not, design process, and configuration. These stages form an analytical theoretical framework providing an abstract picture of BPC and guiding the current research. Altogether, the review provides a comprehensive picture of the current state of business process crowdsourcing.

2.1   The Concept of Crowdsourcing

There is considerable confusion surrounding crowdsourcing terminology in terms of concepts and definitions, as crowdsourcing has continuously developed within different research streams. Thus, it is necessary to explain the concept of crowdsourcing. This section commences with a discussion of the basic ideas behind crowdsourcing. It then compares crowdsourcing with other similar concepts. A definition of crowdsourcing used in this book is then provided.

2.1.1   Main Idea Behind Crowdsourcing

Reliance on the crowd can be traced back to the early 18th century, when the British government decided to provide a cash prize for anyone who could address the problem of precisely calculating ship longitudes (Afuah & Tucci, 2012). Despite a long history of crowd participation, the concept of crowdsourcing has only really emerged in 2006 when Howe (2006b) introduces a process utilising the crowd for fulfilling Internet tasks. This raises the question why crowdsourcing has become so popular only recently. Investigating this question, three main underpinnings behind the emergence of crowdsourcing have been suggested: (1) the crowd, (2) the organisation, and (3) the medium linking the crowd and the organisation. Let us examine these underpinnings.

First, the crowd's wisdom is one of the main underpinnings enable crowdsourcing (Brabham, 2008a; Hosseini, Shahri, Phalp, Taylor, & Ali, 2015b; Saxton et al., 2013). James Surowiecki calls the underpinning as the 'wisdom of crowds', which claimed that "under the right circumstances, groups are remarkably intelligent, and are often smarter than the smartest people in them" (2004, p. xiii). The right circumstances are defined as four prerequisites: cognitive diversity, independence, decentralisation and aggregation. Under these prerequisites, individual ideas in the crowd are not averaged, but aggregated into final solutions. As a result, the aggregated solutions are better than, or at least equal to, the solutions from individual members in the crowd.

Although the wisdom of the crowd is dominant in explaining the concept of crowdsourcing, some extensions should be added to clarify the current capability of crowdsourcing. Malone et al. (2010) extend the underpinnings of crowdsourcing by adding the idea of collective intelligence, which highlights the collective coordination of individuals. This extension opens the solution space of crowdsourcing, based on not only the independence of individuals as the 'wisdom of crowds' but their coordination. Another extension is the ability of crowdsourcing to solve not only single puzzles, but complex tasks that may be decomposed into a large number of simpler tasks (Kittur et al., 2013). As a result, the ability of the crowd should be seen from both its individual and collective intelligence and its capability to manage a large number of tasks.

The second category of underpinnings comes from an organisational viewpoint. As the ability of the crowd seems promising, the next question is whether organisations have any demands for using this ability. In fact, they do. The demands for using external agents to perform tasks has been clearly presented in the management literature: outsourcing (Dibbern, Goles, Hirschheim, & Jayatilaka, 2004), open sourcing (Ågerfalk & Fitzgerald, 2008), and open innovation (Chesbrough, 2013; Seltzer & Mahmoudi, 2013). With outsourcing, organisations have a long history of using contracted resources outside their boundaries. Recently, open sourcing and open innovation have further blurred the organisational boundaries for seeking ideas and innovation beyond the traditional organisational boundaries.

The demands for external sourcing explains the reason why organisations have largely been attracted to crowdsourcing. Organisations utilising the crowd may get benefits similar to outsourcing and open innovation, such as cost saving, customer involvement, and access to outside skills (Rouse, 2010; Saxton et al., 2013). Further, crowdsourcing allows organisations to leverage flexible, on-demand labour. These benefits increase organisational demands for crowdsourcing. It is important to note that although organisational demands to use external resources of crowdsourcing are similar to outsourcing, open sourcing, and open innovation, these concepts are distinctive because of other characteristics, as discussed in the next section.

Given the aforementioned underpinnings, the term 'crowdsourcing' can be etymologically analysed as a combination of two words: *crowd* and *sourcing*. However, the fact that these underpinnings have existed long before the recent emergence of crowdsourcing reveals that another underpinning is needed to enable crowdsourcing. Most of the crowdsourcing literature agrees on the role of the Internet, and in particular the recent dominance of Web 2.0 (Brabham, 2013; Saxton et al., 2013; Zhao & Zhu, 2014). Being globally collaborative, Web 2.0 has changed the nature of online interaction where individuals are no longer passive receivers but active contributors (Brabham, 2013; OReilly, 2007). Brabham (2013) notes that Web 2.0 fastens a voluntary participatory culture onto a global, virtual environment, where Internet users are willing to contribute their skills and labour. Such contributions are perceived as valuable resources for work.

Further, Web 2.0 empowers the open call, which is a distinctive characteristic of crowdsourcing. Because of its millions of users, Web 2.0 extends the scope of the open calls through providing a valuable medium for approaching innumerable anonymous audiences (Saxton et al., 2013). In other words, any given interested participants can now participant in crowdsourcing. It has also eased users to participate in a variety of Internet activities with fewer barriers, e.g. regarding time and space (Brabham, 2013). As a result, it extends the reach and the scope of the crowdsourcing open calls.

This review has shown that, the combination of the crowd, Web 2.0, and organisational demands, can explain the emergence and foundations of crowdsourcing. Given these underpinnings, the IS discipline, which is concerned with people, technologies, and organisations (Bacon & Fitzgerald, 2001), has crowdsourcing as a focus point. This focus point also comes from a strength of the IS research, which draws upon reference disciplines to build its own knowledge base (Baskerville & Myers, 2002). This is exactly the need for the field of crowdsourcing, as a large part of research into crowdsourcing is not very well delimited. All in all, we believe that IS research like the current research can make significant contributions to progress the crowdsourcing field.

The review has also shown that no single underpinning can enable crowdsourcing per se, but rather the combination of the three underpinnings supports the emergence of crowdsourcing. This combination distinguishes crowdsourcing from other concepts, being presented in the next section.

2.1.2 Related Concepts

In another stream of research attempting to clarify the concept of crowdsourcing, many researchers compare this notion with closely related concepts, such as open innovation, outsourcing, open source, and peer production. This section reviews this research stream and discusses crowdsourcing by comparing its similarities and differences with the related concepts.

Among the competing concepts, one often discussed in relation to crowd-sourcing is *open innovation*. Crowdsourcing and open innovation share a common basis where organisations embrace openness to harvest external knowledge and expertise, the opposite of closed innovation. As a result, some researchers suggest that crowdsourcing belongs to or is a technique of open innovation (Marjanovic, Fry, & Chataway, 2012; Seltzer & Mahmoudi, 2013). However, other researchers argue that these two concepts are different, at least in two important points. First, open innovation mainly focuses on innovation processes, while crowdsourcing has been used for various types of tasks (Nakatsu, Grossman, & Iacovou, 2014; Schenk & Guittard, 2011). Second, organisations interact mainly with other firms and their stakeholders in open innovation, but rely on anonymous crowd members in crowdsourcing activities (Flostrand, 2017; Schenk & Guittard, 2009).

Outsourcing is another concept closely related to crowdsourcing. As noted in the previous section, the two concepts are similar on the organisational demands for external agents. As a result, pioneering researchers considered crowdsourcing as a form of outsourcing (Howe, 2006b; Rouse, 2010; Whitla, 2009). Nevertheless, recent conceptualisations of crowdsourcing clearly identify the differences between these two concepts. One major difference is who performs the activities. Actors performing crowdsourcing tasks are informal members of the crowd, while in outsourcing they are mainly established supplier firms. Another difference lies in how to manage these actors. Compared to the official contracts used in outsourcing, crowdsourcing uses an open call where any member in the crowd can participate in the project (Zhao & Zhu, 2014). Finally, financial incentives are the main moti-vation for task performers in outsourcing, whereas crowdsourcing can be based on both intrinsic incentive, e.g. personal enjoyment and hobby, and extrinsic incen-tives, e.g. money (Hossain, 2012; Kaufmann, Schulze, & Veit, 2011; Naderi, 2018).

The literature also distinguishes crowdsourcing from *open source*, although the two concepts are based on resources from the community to accomplish tasks. There are two key aspects distinguishing them: management and engagement. In crowdsourcing, activities are managed by the organisations, whereas in open source these activities are self-managed and community-driven (Brabham, 2013). Regarding to how the community is engaged to perform the activities, crowd-sourcing outcomes can be achieved either independently or collaboratively (Geiger, Seedorf, Schulze, Nickerson, & Schader, 2011), but outcomes from open source are achieved mainly through collaboration. The motivation of the community is another difference between these two concepts. Most of the time, members in open source communities perform tasks based on intrinsic motivation, whereas both intrinsic and extrinsic motivations can be found in crowdsourcing (Kaufmann et al., 2011; Naderi, 2018). Furthermore, unlike open source, crowdsourcing campaigns clearly have intellectual property rights and are not restricted to software development (Wu, Tsai, & Li, 2013).

A few researchers equate crowdsourcing to a form of *peer production* (Mason & Watts, 2009; Wu, Wilkinson, & Huberman, 2009). These researchers believe that peer production sites, like YouTube, can be seen as crowdsourcing because contents on these sites are created by mass individuals in the crowd. However, other researchers argue that crowdsourcing is completely different from peer production. Estellés-Arolas and González-Ladrón-de-Guevara (2012) suggest that crowdsourcing tasks require clear objectives, and thus YouTube, where an individual can upload any video, is not crowdsourcing. In addition, peer production mainly depends on particular communities (Haythornthwaite, 2009; Huberman, Romero, & Wu, 2009), whereas crowdsourcing relies on anonymous members of the crowd, as previously mentioned.

To summarise the above discussion, this review adapts Malone et al.'s (2010) framework to compare crowdsourcing with the related concepts. This framework includes four questions: what needs to be performed, who is performing the task, why people do this, and how the task is being done. An additional question about controlling intellectual property (IP) is added for clarifying the locus of control on the outcomes. By answering the five questions (five rows), Table 2.1 presents the main differences between crowdsourcing and the other concepts. This table reflects that crowdsourcing is a distinctive notion, leading us to investigate the concept per se.

2.1.3 Crowdsourcing Definition

Given the different concepts related to crowdsourcing, we are not surprising that researchers have defined the crowdsourcing concept differently. This section presents a brief history of crowdsourcing definitions in order to understand the concept evolution, and ultimately to form a definition for use in this book.

Until now, crowdsourcing has a short history of one decade. The phenomenon began to appear in 2006 after Howe (2006b) coined this term when he observed several websites utilising Internet users to perform certain activities. It is interesting to note that Howe's (2006b) article has appeared in Wired Magazine—a news media, which indicates that crowdsourcing is a concept spreading from practice to academia. In the article, crowdsourcing was described as the act of organisations through the form of an open call in order to "tap the latent talent of the crowd" (Howe, 2006b, p. 2). In the same year, he proposed the first definition of crowdsourcing.

> Simply defined, crowdsourcing represents the act of a company or institution taking a function once performed by employees and outsourcing it to an undefined (and generally large) network of people in the form of an open call. This can take the form of peer-production (when the job is performed collaboratively), but is also often undertaken by sole individuals. The crucial prerequisite is the use of the open call format and the large network of potential laborers (Howe, 2006a).

Up to now, this definition is among the ones most cited in the field due to its exploratory nature and simplicity. It is worth noting two interesting points from this definition. First, it views organisations as the main caller who operationalise

Table 2.1 Main differences between crowdsourcing and related concepts

	Open innovation	Outsourcing	Open source	Peer production	Crowdsourcing
Tasks	• Only innovation		• Software	• Undefined tasks	• Varied types of tasks • Predefined tasks
Workforce	• Other firms and customers	• Supplier firms	• Software community	• Certain community	• Members of the crowd
Participant motivation		• Extrinsic motivations	• Intrinsic motivations		• Intrinsic and extrinsic motivations
Nature of management and engagement		• Official contracts	• Workflows and quality control managed by community • Collaborative	Collaborative	• Open call • Without official contract • Workflows and quality control mainly managed by the organisations • Collaborative and independent
Control on IP			• IP open		• IP protected

crowdsourcing, which is completely aligned with the promotion of crowdsourcing for organisations in the book. Second, in this definition, crowdsourcing is a sourcing strategy and is an extension of outsourcing.

After 2006, researchers started to explore crowdsourcing and soon published several alternative definitions. Since then, crowdsourcing definitions have evolved over time. Figure 2.2 summarises the evolution of crowdsourcing definitions during the last decade.

After Howe's (2006a) definition, several academic definitions of the concept were published between 2008 and 2009. Extending Howe's (2006a) view, some researchers conceptualised crowdsourcing as a sourcing model where the task performers were the crowd. These researchers further defined who the crowd was and positioned it as a workforce alternative to internal employees and outsourcing agents (Ågerfalk & Fitzgerald, 2008; Howe, 2008; Whitla, 2009; Yang, Adamic, & Ackerman, 2008). At the same time, a parallel approach focused on the intelligence capabilities of crowdsourcing. Researchers in this approach defined crowdsourcing as a problem solving model, where the crowd contributes not only with labour but also with creativity (Brabham, 2008a, 2008b; DiPalantino & Vojnovic, 2009; Vukovic, 2009). As a pioneer researcher in this stream, Brabham (2008b) summarised the notion of crowdsourcing as "a process, a model, for distributed problem solving through the Web" (p. 1). The term 'problem' in Brabham's definition

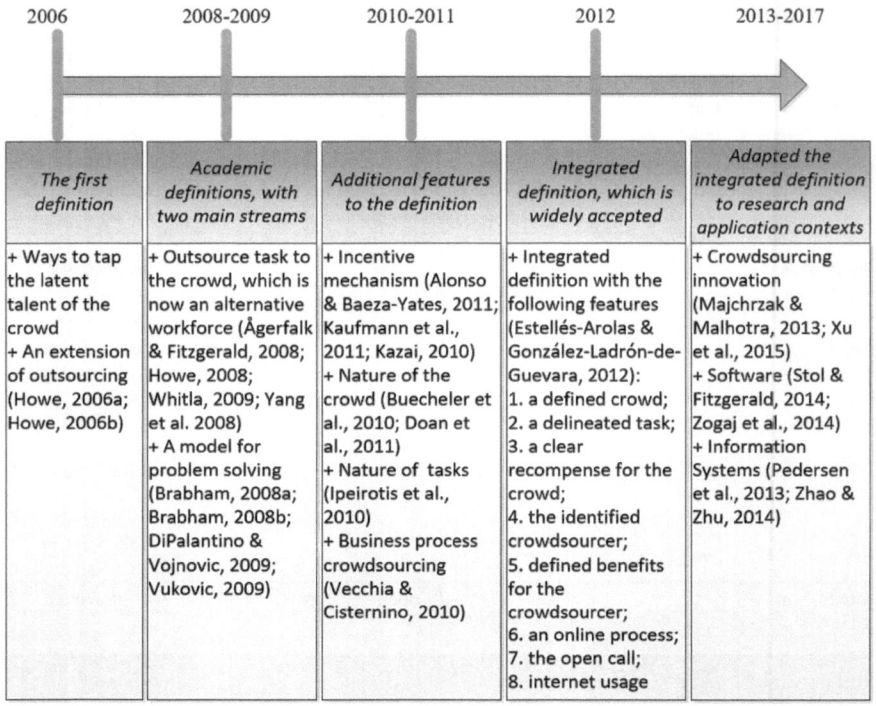

2006	2008-2009	2010-2011	2012	2013-2017
The first definition	Academic definitions, with two main streams	Additional features to the definition	Integrated definition, which is widely accepted	Adapted the integrated definition to research and application contexts
+ Ways to tap the latent talent of the crowd + An extension of outsourcing (Howe, 2006a; Howe, 2006b)	+ Outsource task to the crowd, which is now an alternative workforce (Ågerfalk & Fitzgerald, 2008; Howe, 2008; Whitla, 2009; Yang et al. 2008) + A model for problem solving (Brabham, 2008a; Brabham, 2008b; DiPalantino & Vojnovic, 2009; Vukovic, 2009)	+ Incentive mechanism (Alonso & Baeza-Yates, 2011; Kaufmann et al., 2011; Kazai, 2010) + Nature of the crowd (Buecheler et al., 2010; Doan et al., 2011) + Nature of tasks (Ipeirotis et al., 2010) + Business process crowdsourcing (Vecchia & Cisternino, 2010)	+ Integrated definition with the following features (Estellés-Arolas & González-Ladrón-de-Guevara, 2012): 1. a defined crowd; 2. a delineated task; 3. a clear recompense for the crowd; 4. the identified crowdsourcer; 5. defined benefits for the crowdsourcer; 6. an online process; 7. the open call; 8. internet usage	+ Crowdsourcing innovation (Majchrzak & Malhotra, 2013; Xu et al., 2015) + Software (Stol & Fitzgerald, 2014; Zogaj et al., 2014) + Information Systems (Pedersen et al., 2013; Zhao & Zhu, 2014)

Fig. 2.2 Evolution of crowdsourcing definitions

should be understood in a broad sense, including not only R&D problems but also design and innovation.

During 2010–2011, there was a boom of crowdsourcing definitions, aligning to a diverse set of practices and an increasing number of research interests in the field. At this stage, researchers adopted different theoretical bases and models to investigate several aspects of crowdsourcing. Depending on the research foci, the related features were depicted and added to crowdsourcing definitions, including the nature of the crowd (Buecheler, Sieg, Füchslin, & Pfeifer, 2010; Doan, Ramakrishnan, & Halevy, 2011), the nature of tasks (Ipeirotis, Provost, & Wang, 2010), and incentive mechanisms (Alonso & Baeza-Yates, 2011; Kaufmann et al., 2011; Kazai, 2010). These definitions, on the one hand, contribute to clarifying several features of the concept. On the other hand, definitions with too many additional features suffer from diversity and sometimes conflict with each other, which makes crowdsourcing hard to comprehend.

Addressing this problem, in 2012, Estellés-Arolas and González-Ladrón-de-Guevara (2012) aimed at establishing an integrated definition of crowdsourcing. Driving their research was the goal to conceptualise 'any given crowdsourcing activity' by reviewing the diverse definitions extracted from literature. The authors selected 209 crowdsourcing articles and analysed 40 of them that present original definitions of crowdsourcing. The results suggest eight key characteristics of crowdsourcing: a clearly defined crowd, a task with a clear goal, a clear recompense for the crowd, an identified crowdsourcer (or caller), defined compensation for the crowdsourcer, an online process, an open call, and Internet usage. The authors then integrate these characteristics into a single comprehensive definition.

> Crowdsourcing is a type of participative online activity in which an individual, an institution, a non-profit organisation, or company proposes to a group of individuals of varying knowledge, heterogeneity, and number, via a flexible open call, the voluntary undertaking of a task. The undertaking of the task, of variable complexity and modularity, and in which the crowd should participate bringing their work, money, knowledge and/or experience, always entails mutual benefit. The user will receive the satisfaction of a given type of need, be it economic, social recognition, self-esteem, or the development of individual skills, while the crowdsourcer will obtain and utilize to their advantage what the user has brought to the venture, whose form will depend on the type of activity undertaken (Estellés-Arolas & González-Ladrón-de-Guevara, 2012, p. 197).

Due to its comprehensiveness, this definition has been widely accepted and frequently referred to. Yet, the definition is wordy and thus complex, which decreases its practical use. As a result, many recent studies have implicitly or explicitly adapted the aforementioned eight characteristics by simplifying and adjusting them to their own research and application contexts. For instance, crowdsourcing software emphasises the software tasks being crowdsourced (Stol & Fitzgerald, 2014) and the intermediated platforms (Zogaj, Bretschneider, & Leimeister, 2014); crowdsourcing innovation focuses on the innovative ability of the crowd (Majchrzak & Malhotra, 2013; Xu, Ribeiro-Soriano, & Gonzalez-Garcia, 2015). These adaptations show that there is no unique universal definition of crowdsourcing appropriate for all applications and research contexts, though Estellés-Arolas and González-Ladrón-de-Guevara's definition may form a basic understanding.

Aligning to the most recent trend, this book simplifies and adapts the definition by Estellés-Arolas and González-Ladrón-de-Guevara (2012) to the organisational context. We use the following definition.

> Crowdsourcing is an online strategy in which an organisation proposes defined task(s) to the members of the crowd via a flexible open call. By undertaking the task(s), the members contribute their work, knowledge, skills and/or experience and receive rewards, including economic rewards, social recognition, self-esteem, or the development of individual skills. The organisation will obtain contributions from the crowd and will utilise the results to meet business goals.

To sum up, this review has shown the conceptualisation of crowdsourcing, through three facets. The first facet has shown the three main pillars of crowdsourcing: the organisational demands for external sourcing; the ability of the crowd; and the intermediary Web 2.0. These pillars together enable crowdsourcing. The second facet has compared and differentiated crowdsourcing with related concepts, like open innovation, outsourcing, open sourcing and peer production. It emphasises the distinctive characteristics of crowdsourcing. The final facet has shown a brief evolution in crowdsourcing definitions. It then proposes the definition that to be used in this book. From these facets, it is worth noting that although a few referencing theories have been applied to explain crowdsourcing, like the wisdom of the crowd, open innovation, and outsourcing practices, the distinctive characteristics of crowdsourcing state that these theories are not predominant in the phenomenon. Consequently, crowdsourcing is a concept per se that needs its own structures and establishment.

2.2 Crowdsourcing Classifications, Taxonomies, and Typologies

As classifications and taxonomies are useful to structure knowledge in the IS discipline (Nickerson et al., 2012), an extensive body of crowdsourcing literature is devoted to crowdsourcing classifications, taxonomies, and typologies. Although they contribute to structuring the domain, each of them focuses on different crowdsourcing elements. This section considers the popular classified elements: applications, tasks, members and platforms, which are essential for exploring crowdsourcing processes. In particular, this section aims to answer the following four questions: What are the crowdsourcing applications? Which types of tasks can be crowdsourced? Who will perform these tasks? And where can these tasks be performed?

Before proceeding, it is useful to clarify terminologies of classification, as a variety of them have been used in the literature, including classification, taxonomy, and typology. The term *classification* has been used as both a product and a process of classifying objects according to a particular system (Fettke & Loos, 2003). *Taxonomy* and *typology* are two forms of classification that usually deploy multi-dimensions to classify objects into categories. Some researchers further distinguish taxonomies as empirical classifications and typologies as conceptual classifications (Bailey, 1994). However, other researchers suggest using classification, taxonomy, and typology interchangeably (Gregor, 2006; Nickerson et al., 2012). We follow this suggestion as we observe that the crowdsourcing literature commonly refers to the three terms in an interchangeable way. Consequently, this book uses these terminologies more or less synonymously.

2.2.1 Applications

Crowdsourcing has been applied to different *applications*. Howe (2006b) discusses the crowdsourcing concept through several applications in solving real business problems, including InnoCentive for problem solving, iStockphoto for image exchange, and AMT for micro tasks. In addition to business applications, crowdsourcing can also be applied to scientific research, urban planning, public health, and cultural heritage. Given the variety of crowdsourcing applications, their classifications are necessary for understanding the potential utility of crowdsourcing. Addressing this necessity, several application classifications have been proposed in the literature, which are now reviewed in detail. The review moves from simple classifications, defining for specific areas, to more inclusive typologies at the level of work practice.

Some studies, maybe for simplification, classify crowdsourcing applications specifically for one single area. For instance, Whitla (2009), focusing on marketing-related areas, classifies crowdsourcing applications into three function categories, namely marketing research, product development, and advertising and promotion. Gomes et al. (2012) propose a crowdsourcing taxonomy with a focus on the context of musical productions. Based on what crowdsourcing can be utilised for, the taxonomy identifies six types of applications: music co-creation, decision support, crowdsourced music collection and management, promoting music information, market place, and crowd funding.

Adopting a broader approach, other studies propose a number of application typologies that can be used in multiple domains. Kleeman et al. (2008) explored start-up crowdsourcing applications, and typologically grouped them according to their functions. As a result, seven application types are defined, namely product development and configuration, product design, permanent open calls, competitive bids, community reporting, product rating, and customer-to-customer support. This approach is also employed by Brabham in his recent book, *Crowdsourcing* (2013). He, surveying crowdsourcing cases, conceptualises them into four different functions, including knowledge discovery and management, broadcast search, peer-vetted creative production, and human intelligence tasks. Other typologies which follow a similar approach are mentioned in the literature (Man-Ching, King, & Kwong-Sak, 2011).

The studies reviewed so far have a common point. They suggest that *function* is a main dimension to classify crowdsourcing applications. Agreeing with this suggestion, we note, however, that functions alone seem not enough, since a *context*, where crowdsourcing is applied, plays an equally significant role. Chandler and Kapelner (2013), who conducted an experiment on AMT, find that if the context is explained, more workers are willing to participate in the crowdsourcing application. In addition, whether it is a business or non-business context strongly influences the application operation because the context directly links to incentives that may be required to attract people to participate in the crowdsourcing applications (Rosen, 2011).

Given the importance of contexts in characterising crowdsourcing applications, two dimensions: *function and context* together are likely more appropriate to classify applications. This appropriateness is supported by Zhao and Zhu (2014), who broadly reviewed crowdsourcing applications. By deductively analysing 126 applications, they propose a typology based on the two dimensions of function and context. In the first dimension, these authors group functions into four categories: design and development, test and evaluation, idea and consultant, and others. In the second dimension, two categories of contexts are suggested: business and non-business. A business context consists of for-profit organisations, while

Table 2.2 Typology of crowdsourcing applications (Zhao & Zhu, 2014)

Context	Function			
	Design and development	Idea and consultant	Test and evaluation	Other
Business	• Threadless • IStockphoto	• MyStarbucks Idea • InnoCentive	• Crowdspirit	• AMT
Non-business	• NextStopDesign (Brabham, 2012)	• QuestVille	• UTest	• Wikipedia

non-business includes non-profit organisations and institutions (Zhao & Zhu, 2014). Although this dimension considers contexts at an organisational level, we suggest the level of application is more precise for this dimension. The reason is that one organisation may have both business and non-business applications, such as Amazon owning AMT for profit and QuestVille for non-profit (Saxton et al., 2013). In this case, the context dimension does not associate with the organisation but with its applications. Therefore, this book adopts the typology proposed by Zhao and Zhu (2014), yet considers both the function and context dimensions from the viewpoint of crowdsourcing applications (Table 2.2).

2.2.2 Tasks

Tasks are basic elements of a crowdsourcing application. Organisations define tasks and send them to members in the crowd, who will perform these tasks. Several studies have suggested clearly identifying task characteristics before crowdsourcing, which helps to determine the appropriate approach for a particular task (Malone et al., 2010; Nakatsu et al., 2014; Rosen, 2011). Several taxonomies characterising tasks have been proposed in the literature.

There are two main views on building task taxonomies regarding whether tasks should be examined in related with other elements or by its own nature. On the one hand, a number of published taxonomies are based not only on task properties, but also on "key questions [elements] associated with a single task" (Malone et al., 2010, p. 22). Rouse (2010) provides one the of the earliest taxonomies, structured around three dimensions: nature of the task, distribution of benefits, and forms of motivation. In a similar vein, Malone et al. (2010) propose a multi-dimensional classification after analysing 250 instances of crowdsourcing. The classification is based on four basic questions: what is being crowdsourced, who is performing the task, why would people do this, and how is the task to be done. In these cases, the developed taxonomies suggest multiple dimensions for classification, with task as a central dimension.

On the other hand, another group of published taxonomies classifies tasks by their own nature. By examining the task characteristics in practical applications,

Table 2.3 Examples of crowdsourcing task types (adapted from Schenk and Guittard (2011))

Complexity	Participation mode	
	Individual (integrative)	Competitive (selective)
Simple	**Market place** • Simple tasks (MicroWorkers, AMT & Taskcn)	**Simple contest** • Answering simple questions (Ask Ville by Amazon & Yahoo Answers)
Skilled	**Collective intelligence** • Writing & editing (Wikipedia) • Writing academic papers (Tomlinson et al., 2012)	**Problem solving contest** • Designing T-shirts (Threadless) • Problem solving (InnoCentive)

Schenk and Guittard (2009, 2011) suggest two dimensions to classify crowd-sourcing tasks. The first dimension classifies tasks as simple, complex or creative. Simple tasks are jobs that can be performed without any specific skills, such as text transcription. Complex tasks require expertise and skills, such as problem solving. Creative tasks relate to individual creativity, such as logo design. The second dimension distinguishes between the integrative and selective nature of tasks (Schenk & Guittard, 2011). Other taxonomies in this group can also be found in work by Nakatsu et al. (2014).

Given the existing taxonomies, a critical question is which one will be used in this book. To answer this question, the book adopts Nickerson et al.'s (2012) suggestion that usefulness is the key criterion to evaluate a taxonomy and its dimensions. Thus, choosing dimensions for task classification in the book should be based on their usefulness for the research focus. That is, the establishment of BPC, consisting of three stages: the decision to crowdsource, process design, and configuration, will be discussed in Sect. 2.3.3. In the first stage, the complexity of tasks plays a role in the decision to crowdsource (Zhao & Zhu, 2014). In the remaining stages, whether tasks are achieved individually or competitively, influences the crowdsourcing design and operation, because it directly affects how the tasks should be planned, coordinated, and performed.

Consequently, this study adapts the two dimensions proposed by Schenk and Guittard (2011): *task complexity* (simple and skilled) and *the difference between integration and selection based crowdsourcing*. Table 2.3 presents examples of different types of crowdsourcing tasks (and their related platforms).

2.2.3 Members of the Crowd

Crowd members are actors who accomplish tasks in crowdsourcing applications. There are several studies examining characteristics of crowd members. In general, these studies can be grouped into one of two research directions. The first direction examines the crowd characteristics by exploring its properties, such as who members of the crowd are and where they come from. Studies by Mason and Suri

(2012) and Brabham (2011) can be categorised in this direction. Another direction studies the crowd as a whole and assesses its performance (Chandler & Kapelner, 2013; Stewart, Lubensky, & Huerta, 2010).

In the first direction, Brabham (2011) changed the popular image of the crowd being amateur. By conducting a survey on iStockphoto and several interviews on Threadless, he finds that members on both of these platforms "seem ill-fitted to the amateur label" (Brabham, 2011, p. 399). Specifically, 47% of participants on IStockphoto described themselves as professional, while many members on Threadless have previously performed real design activities (Brabham, 2011). The argument that the crowd is not wholly amateur, and thus can be in competition with professionals, is also supported by other studies. Jeppesen and Lakhani (2010), who examined the members on InnoCentive, report that "65% of solvers reported holding Ph.D. degrees" (p. 1026). Poetz and Schreier (2012), conducting a case study in the idea contest, find that the crowd can outperform the professionals in certain aspects of idea quality.

Similar to Brabham (2011) in exploring the properties of the crowd, Mason and Suri (2012) focusing on AMT present several aspects of AMT's 'workers'. For instance, there are about 100,000 workers on AMT, who are mainly from USA and India. This crowd has more females than males. These characteristics are consistent with findings from another study of 1,000 workers using the same platform conducted by Paolacci et al. (2010), who further report that the hourly average wage on this platform is $1.66. From these observations, three reasons provided by Mason and Suri (2012) to choose AMT for online experimentation can be generalised as the crowd characteristics on AMT: large pool of workers, pool diversity and low cost.

In the other direction, studies investigating performance of the crowd as a whole show that the performance is not as promising as the characteristics presented in the first direction. The fact that not all members of the crowd actively performed tasks was analysed by Stewart et al. (2010), building on the participation inequality rule of online community (Nielsen, 2006). By analysing a crowd of 400,000 members in a language translation application, these authors separate members of the crowd into three categories: super contributors (1%) who provide the most contributions, contributors (66%) who provide the moderate contributions and outliers who rarely contribute (33%). Further analysing the crowd members, Kazai et al. (2011) find that members may perform tasks dishonestly, randomly, or sloppily. In a similar vein, Vuurens and de Vries (2012) suggest a theoretical typology classifying four types of workers regarding their behaviours: diligent workers, sloppy workers, random spammers, and uniform spammers.

From the given discussion, some characteristics of the crowd should be highlighted. On the one hand, the crowd is promising in terms of providing a large, diverse, and low-cost workforce (Mason & Suri, 2012). It may also include 'self-selected' experts (Brabham, 2011). On the other hand, members of the crowd have different levels of contribution for accomplishing tasks (Stewart et al., 2010; Vuurens & De Vries, 2012). We note that the reviewed studies mainly identify the crowd characteristics based on individual applications and platforms, which implies that the characteristics of the crowd may be different in varied applications and contexts.

2.2.4   Platforms

Platform is another key element of crowdsourcing, which serves as a mediator connecting the organisation and the crowd (Hirth, Hoßfeld, & Tran-Gia, 2011). Vukovic (2009) describes several functions of a crowdsourcing platform: "issues authentication credentials for requestors and providers when they join the platform, stores details about skill-set, history of completed requests, handles charging and payments, and manages platform misuse" (p. 687). Aiming to utilise crowd-sourcing, organisations can choose either to develop their own platforms or to use the available ones provided by a third party. Each approach has its own advantages and disadvantages.

Some examples of organisations developing their own crowdsourcing platforms are Threadless, and MyStarbucksIdea in the business context, and Next Stop Design in the non-business context (Brabham, 2012). Through self-development, organisations can fully control the application and its functions, such as tracking geographic locations of visitors for research purposes in case of Next Stop Design (Brabham, 2012). Another advantage of this approach is building closer relation-ships with their own customers, who associate with the platforms. For instance, Threadless uses a self-developed platform to ask customers to design T-shirts, and then sells those T-shirts to the customers (Brabham, 2010). Despite these advan-tages, this approach requires organisations having experts and experience in developing crowdsourcing platforms, since the platform development may have several complex requirements (Adepetu, Ahmed, Al Abd, Al Zaabi, & Svetinovic, 2012; Vukovic, 2009).

As an alternative to self-development, organisations can hire existing crowd-sourcing platforms built by a third party to deploy their applications. The existing platforms can be further divided into two kinds: specialised platforms, which focus on particular tasks (Hirth et al., 2011; Hoßfeld et al., 2013); and horizontal plat-forms, which can be utilised for different types of tasks (Kucherbaev et al., 2013). Two examples of a specialised platform are InnoCentive that utilise the crowd only for problem solving purposes (Malone et al., 2010), and TopCoder that uses crowdsourcing for software engineering (Mao, Capra, Harman, & Jia, 2017). Differently, horizontal platforms publish different types of tasks. AMT is a typical horizontal platform, which can help an organisation to do several tasks, including data collection, transcription, and image categorisation. To configure a crowd-sourcing application on horizontal platforms like AMT, organisations need to use the provided application programming interface (API) (Ipeirotis et al., 2010). Thus, basic programming skills and platform knowledge are required.

Using existing platforms can save organisations' resources, which would otherwise need to be spent on developing their own new platform. Furthermore, existing platforms already have their own members, who are available for new crowdsourcing applications. For instance, an application developed on AMT can use any number of 100,000 available workers (Mason & Suri, 2012). However, existing platforms limit crowdsourcing applications to what is supported by the platforms. From the above discussion, it is important to note that both approaches have their own pros and cons, which should be considered when making the decision to build or to hire a crowdsourcing platform. Table 2.4 summarises the main pros and cons of the discussed platforms types: self-development, specialised platforms, and horizontal platforms.

In summary, the preceding review identified major classifications in the crowdsourcing literature, including applications, tasks, members, and platforms. On the one hand, these classifications suggest possible options and features that are available in crowdsourcing, which contributes to initially structure the domain. On the other hand, many of them have focused on specific aspects of crowdsourcing and on specific crowdsourcing contexts. This leads to differences, sometimes conflicting, on the domain structures. For instance, crowdsourcing tasks can be classified differently using either four dimensions (Malone et al., 2010), three dimensions (Rouse, 2010), or two dimensions (Schenk & Guittard, 2011).

We believe that this is symptomatic of a more general issue with the ad hoc focus of the existing classifications. That is, the domain is structured through its individual elements without synthesis and coordination between them. If we cannot address this ad hoc issue, and if new studies continue to propose crowdsourcing taxonomies that are solely relevant to specific elements, the domain may end up with ambiguity over its structure. Given that, there is a strong need for a more comprehensive integrated approach in order to structure the domain.

Addressing the need, we suggest that a domain ontology and a process view are necessary for structuring the domain. Regarding the former, a domain ontology

Table 2.4 Crowdsourcing platform types

Dimension	Self-development platforms	Platforms by a third party	
		Specialised platforms	Horizontal platforms
Control	Fully control	Depending on platform	
Customer relationship	High	Low	
Development effort	High	Low	
Tasks being crowdsourced	Organisational focus	Platform focus	Diversity
Availability of crowd	Low	Medium	High
Crowd expertise	High	High	Low
Examples of platforms	• MyStarbucksIdea • Next stop design (Brabham, 2012)	• InnoCentive • TopCoder	• AMT • Microworkers

enables us to integrate the existing classifications. Nickerson et al. (2012) suggest that ontologies are the next stage of taxonomy development. Further, Corcho et al. (2003) highlight ontologies for their comprehensiveness and ability to structure domain knowledge. Regarding the need for a process view, we note that the existing classifications have not been linked together yet, which is necessary to constitute the whole crowdsourcing practice. This highlights the process view connecting individual elements in a meaningful way. This process view is a central of the book, where we address crowdsourcing processes and BPC, and is the focus of the next section.

2.3   Current State of Business Process Crowdsourcing

This section aims to paint an overall picture regarding the emerging state of business process crowdsourcing (BPC). The section starts with describing crowdsourcing processes, an antecedent of BPC. It then provides a review of BPC related literature, followed by an initial conceptualisation and a theoretical framework of BPC. By channelling the related research, the framework guides the current research and paints an abstract picture of BPC.

2.3.1   Crowdsourcing Process

The notion of a *crowdsourcing process* is critical to operationalise a crowdsourcing strategy. Thus, it is a recurrent topic in the crowdsourcing literature. We use the term 'process' to refers to a set of systematic activities to complete some deliberate results. Well-coordinated processes are assumed not only to generate better crowdsourcing results (Thuan et al., 2017), but also to deploy crowdsourcing applications more efficient and with less cost (Tranquillini et al., 2015). Numerous studies have devoted attention to the topic. By and large, existing studies on crowdsourcing processes can be classified into two basic genres according to its view: high and low levels of granularity.

With high level of granularity, some studies adopt a holistic view to conceptualise the crowdsourcing process. Early, research referred to crowdsourcing processes with an understanding purpose. Consequently, crowdsourcing processes were conceptualised by rich descriptions with several illustrative examples (Leimeister, Huber, Bretschneider, & Krcmar, 2009; Whitla, 2009), and by identification and description of actions executed by different crowdsourcing actors (Geiger et al., 2011; Vukovic, 2009; Wexler, 2011). At this early time, crowdsourcing processes were mostly studied together with other foci like crowdsourcing applications and taxonomies, rather than as a separate primary research focus. Before moving to review studies that primarily investigate crowdsourcing processes, we synthesise the existing descriptions to provide a narrative sketch of the crowdsourcing process. More precisely, we adapt the earliest but most widely used description by Whitla (2009) and add into it supplementary descriptions. As a result, a crowdsourcing process can be described as follows.

The crowdsourcing process starts with a go/no-go decision whether to choose crowdsourcing to perform the organisational tasks or not (Thuan, Antunes, & Johnstone, 2013; Wexler, 2011). If the decision to crowdsource is made, the organisation then creates an open call to release the defined tasks to the crowd. This step is normally done through a platform developed by either a third party or the organisation itself. Through the open call, the organisation approaches members of the crowd, who can belong specifically to a particular community or just anyone willing to complete the task. The members accomplish these tasks individually or collaboratively, and then submit the results back to the organisation, which assesses the quality of the results. Incentives will be given to the members if the organisation is satisfied with the submission results (Whitla, 2009). The results are intended to be incorporated into organisational activities (Leimeister et al., 2009; Wexler, 2011).

Keeping in mind the initial descriptions, researchers started to explore crowd-sourcing processes from a high level of granularity. Aiming to identify the main structures of the process, they commonly adopted an abstract view to discover the main stages and concerns in the process. Brabham (2009, 2012), exploring a crowdsourcing project for public participation in transit planning, formulates a crowdsourcing process using four stages. First, a problem that needs to be solved and its related information are clarified. Second, an open call is sent to the crowd through a self-developed website. This call includes data necessary to solve the problem, reward information and the intended format of the solutions. Third, crowd members can choose to participate in the project. Finally, the organisation evaluates the proposed solutions to choose the winners.

Also adopting an abstract broad view, Muhdi et al. (2011) conducted an explorative case study to analyse twelve crowdsourcing projects. As a result, they formulate the main operations in the crowdsourcing process as five stages: delib-eration, preparation, execution, assessment, and post-processing. In the first stage, organisations analyse crowdsourcing and "decide whether the crowdsourcing approach is appropriate to solve their internal problem[s]" (p. 322). If the decision to crowdsource is made, the second stage involves choosing a particular platform that is appropriate for the crowdsourcing activity. The next two stages are dedicated to executing the crowdsourcing activity on the chosen platform, and evaluating the received results. The final stage transfers the received results, such as ideas and solutions, to real organisational implementation.

In a similar vein, Stol and Fitzgerald (2014) conducting case study research recently examined crowdsourcing processes in the context of software companies. However, they structure their findings differently compared to the two aforemen-tioned studies. More precisely, instead of formulating crowdsourcing processes as a set of sequential stages, they identify major building blocks of the crowdsourcing process, including task decomposition, coordination and communication, planning and scheduling, quality assurance, managing knowledge and intellectual property, and providing incentives to the crowd. Similar approaches that formulate main elements of the crowdsourcing process by case study are quite common (Ågerfalk, Fitzgerald, & Stol, 2015; Zogaj et al., 2014).

Overall, this group of studies views the crowdsourcing process as an important research focus and contributes empirical efforts to formulate the main stages and building blocks that comprise the crowdsourcing process. However, the main research methods adopted in these studies are exploratory case studies (Ågerfalk et al., 2015; Muhdi et al., 2011; Stol & Fitzgerald, 2014; Sutherlin, 2013). The exploratory nature, together with the particular investigated cases/contexts, explains the existence of different, likely one-off crowdsourcing processes. Furthermore, as this group of studies target to provide an overall picture of the crowdsourcing process, they focus on high-level abstract concepts and thus face significant gaps mapping the abstract concepts to specific workflows or activities, necessary to establish the crowdsourcing process.

With low level of granularity, a large number of studies have investigated varied aspects of the crowdsourcing process. Although they have helped specify workflows and activities necessary to establish the crowdsourcing process, their ad hoc nature has been repeatedly complained (Geiger & Schader, 2014; Zhao & Zhu, 2014). This ad hoc nature is further revealed through two aspects. First, different research methods have been adopted to examine the specific activities of the crowdsourcing process. For instance, methods for researching task definition include lab experiments (Khazankin et al., 2012a), open-ended and quantitative surveys (Schulze, Seedorf, Geiger, Kaufmann, & Schader, 2011), and engineering design (Bozzon, Brambilla, Ceri, & Mauri, 2013). These differences contribute to clarifying different aspects of the activity, yet a comprehensive approach is still missing. Second, the domain is lacking a strong knowledge base guiding crowdsourcing process establishment (Palacios, Martinez-Corral, Nisar, & Grijalvo, 2016; Zhao & Zhu, 2014). As a result, the domain knowledge remains scattered, varied and sometimes conflicting.

Given the existence of the large number of studies in this group, this section does not intend to review them one by one, which will be the focus of the scoping knowledge source in Sect. 3.1. Rather, we summarise other major literature reviews, which characterise the complexity and isolated concerns of the crowdsourcing field. Among a few literature reviews in the domain, we focus on the two most recent and major reviews.

In 2014, the first major review was published by Zhao and Zhu (2014). These authors identified 55 crowdsourcing papers, based on a systematic search and selection of all major scholar databases in the period from 2006 to 2011. Analysing the papers, they suggest that "empirical studies have been conducted almost entirely on events/processes" (p. 419). These authors further map these ad hoc foci into major themes, and outline future research directions, including motivation to participate, participant's behaviour, making the decision to adopt crowdsourcing, governance and implementation, quality control and evaluation, incentive mechanisms, and technological issues, which are all major topics of crowdsourcing process studies. The review also indicates the emerging nature of the domain because only a small part of the studies (16%) is based on theoretical foundations.

Amrollahi (2015), among the most recent reviewers, aims at synthesising the crowdsourcing literature into a process model. He started by searching crowdsourcing papers in the period of 2009 to early 2014, coming up with 566 papers, and then selected 39 papers directly focusing on the crowdsourcing process. The review contributes to a better understanding of the crowdsourcing process in three ways. First, it proposes a process model to structure the crowdsourcing process. To an extent, the model is more or less aligned with the stages of the crowdsourcing process described in the aforementioned review. Second, the review indicates a strong development of the field, with a significant increase in the number of papers published recently (566 papers). Lastly, Amrollahi (2015), aligning with Zhao and Zhu, concludes the ad hoc feature of the current literature, and further highlights that crowdsourcing process research remains scarce, with only 39 related papers that can be identified out of the 566 papers found.

In summary, the crowdsourcing processes have been studied from both high and low levels of granularity. With high level of granularity, some studies choose an abstract conceptualisation when exploring a crowdsourcing process. As a result, these studies identify main stages and issues that should be considered in the crowdsourcing process. They contribute to the structures of the crowdsourcing process, which enable us to incorporate an analytical framework discussed in the next major section of this review. However, it is important to note that these studies are more focused on highly abstract conceptual understanding and thus detailed activities are still missing.

With low level of granularity, a larger number of studies examine individual processes/events from varied deconstructed aspects. They provide various contributions, reported in case studies, expert opinions, usability studies, experiences, and other engineering development. Though realising the importance of the high-level view, their investigation tends to focus only on parts of the process (Thuan et al., 2017). The ad hoc nature of these studies is repeatedly complained and is highlighted by the two major reviews in the domain. Furthermore, these reviews highlight that research into the crowdsourcing process as a whole is scant, something also suggested by others (Hossain, Kauranen, & Busi, 2015; Mao et al., 2017). As a result, the domain is still unstructured and lacks "a comprehensive guideline through which practitioners can initiate and manage their crowdsourcing projects" (Amrollahi, 2015, p. 2).

To conclude, a few studies cover the crowdsourcing process as a whole without its parts, while a large number of studies investigate the concept through its parts without the whole. The domain is characterised by a large number of ad hoc knowledge sources, which are scattered, varied and sometimes conflicting. This indicates the lack of a solid knowledge base founding the crowdsourcing process. What is also missing is an integrated view of the two levels of granularity, which can provide a complete picture on decomposed activities of the crowdsourcing process and their coordination.

Such an integrated view can be achieved through a business process lens, which has rarely been adopted in the crowdsourcing field. This points us to the concept of business process crowdsourcing, conceptualised in the following section.

2.3.2   *Business Process Crowdsourcing*

This section explores business process crowdsourcing (BPC). As this concept is relatively new, the review is limited to a small amount of existing relevant literature.

Based on the need for an integrated picture of crowdsourcing processes, this book investigates crowdsourcing using a business process lens. We refer this view as *Business Process Crowdsourcing* (BPC). The term BPC was first coined by La Vecchia and Cisternino (2010) to describe a model allowing organisations to utilise the power of the crowd for their internal business processes, as an alternative to Business Process Outsourcing (BPC). We further define the concept as a way to use crowdsourcing as repeatable organisational business processes. The etymology of the BPC concept is a combination of the phrase *business process* with the word *crowdsourcing* (Thuan et al., 2017). We bring the concept of business process into the concept of crowdsourcing, and consider them as equally important. As the concept of crowdsourcing has been extensive discussed in this book, here we discuss the concept of business process. A business process, according to van der Aalst and Hee (2004), is defined as a combination of individual activities and a workflow describing their logical order. A business process serves as a template for creating multiple, real life instances of the same process, which organisations may create repeatedly and concurrently.

> Given that, this book defines *BPC as a set of activities completed by crowdsourcing entities, in conjunction with a logical coordination of these activities, that collectively form the entire business process.*

Our proposition is that BPC proposes an efficient structured approach for organisations to establish a crowdsourcing process. This efficiency is realised through three roles. First, BPC can help establish repeatable crowdsourcing processes. Inheriting from the business process construct, BPC serves as a template for which organisations create multiple instances of the same repeatable crowdsourcing process. The repeatable characteristic enables analysis of individual aspects of crowdsourcing and their coordination into an organisational workflow (La Vecchia & Cisternino, 2010; Lüttgens et al., 2014). By establishing well-organised workflows, organisations can integrate the crowdsourcing strategy with their day-to-day business processes (Tranquillini et al., 2015). Thus, it enables the incorporation of the crowdsourcing capabilities into the organisational value proposition.

Second, with BPC organisations can start standardising crowdsourcing processes. A pre-condition for process standardisation is that we can comprehend all related activities and their relationships (Thuan et al., 2017). Relying on both the individual and coordinated views, BPC is in a unique position for this comprehension. More precisely, BPC can provide both a detailed view to understand the deconstructed aspects, and a holistic view to understand their relationships, both necessary for process standardisation. This is similar to the role of the business process view on standardisation of outsourcing (Wüllenweber, Beimborn, Weitzel, & König, 2008).

Finally, BPC contributes to move crowdsourcing toward a more well-defined status. The current ad hoc status of the domain has been noted and discussed in the previous section of the review. Bringing a business process lens to crowdsourcing, BPC allows analysing and defining the basic workflows of crowdsourcing processes, and enabling us to build crowdsourcing processes on top of existing business process management (BPM) technology (Khazankin et al., 2012a; Satzger, Psaier, Schall, & Dustdar, 2011; Tranquillini et al., 2015). In this sense, BPC is expected to efficiently establish crowdsourcing as a common well-defined practice.

Given these important roles, BPC has recently attracted considerable research attention. Many researchers have called for further research on BPC, especially how to conceptualise, establish, and coordinate it. Vukovic et al. (2010) raise "how does crowdsourcing become an extension of the existing business process" (p. 7). Khazankin et al. (2012a) echo similar question and complain about "the lack of an integrated way to execute business processes based on a crowdsourcing [platform]" (p. 1). Similarly, other studies have recently highlighted the demand to build a dedicated crowdsourcing process. This demand increases when organisations have recently used crowdsourcing for core organisational processes like product development (Djelassi & Decoopman, 2013), innovation processes (Lüttgens et al., 2014), industrial processes (Muntés-Mulero et al., 2013), and software development processes (Stol et al., 2017), which have to be coordinated with other organisational business processes.

In spite of these calls, there has been little investigation into BPC and thus how to establish BPC has not been fully examined in the literature. Some prior studies have touched different aspects of BPC. Satzger et al. (2011) seek to help organisations "fully automate[d] deployment of their tasks to a crowd, just as in common business process models" (p. 67), but focus only on choosing suitable workers to perform tasks. Similarly, Khazankin et al. (2012a) highlight the need for organising business processes based on crowdsourcing, but they investigate only a part of the problem, which is how to optimise task properties for supporting business process execution.

A few recent models/frameworks conceptualising crowdsourcing processes contribute to the understanding of BPC. One of the earliest model is proposed by Pedersen et al. (2013). From a process perspective, they in-depth analysed existing research in the domain for conceptualising crowdsourcing. As a result, they propose a conceptual model, organised as an Input-Process-Output structure. The model explains key dimensions of crowdsourcing, including problems, technology, processes, governance, people, and outcomes, which provides a starting point for further study on crowdsourcing processes.

Also analysing existing research in the domain, Hetmank (2013, 2014) aimed at understanding crowdsourcing systems and their components. For this purpose, he suggests a model comprising of four components: user management, task management, contribution management and workflow management (Hetmank, 2013). Based on the identified components, Hetmank (2014) further proposes a lightweight ontology defining vocabularies of crowdsourcing systems. The vocabularies specify classes and properties, which are useful for crowdsourcing system development. Yet, further evaluation is needed to empirically test the proposition before its practical use, as noted by the author (Hetmank, 2014).

The crowdsourcing process has also been modelled using BPM technology. Tranquillini et al. (2015), based on Business Process Model and Notation (BPMN) technology, modelled workflow patterns of the crowdsourcing processes. They also designed a run-time environment operating these patterns in order to support the workflow enactment. As a result, the study offers a modelling language supporting crowdsourcing workflow enactment and a visual editor that allows organisations to graphically create and manage their crowdsourcing processes. We note that although this work can enact, prototype, and configure a crowdsourcing process, it can only maximise its contribution with the assumption that organisations have already had clear structures of the crowdsourcing process. In other words, this work provides useful supports to configure business process crowdsourcing, which can only be possible if BPC can be clearly established. This further highlights the role of BPC establishment.

Overall, since crowdsourcing needs to evolve from an ad hoc one-off process, we bring the business process lens to research crowdsourcing. We have introduced the concept of BPC and described its possible roles in moving crowdsourcing processes forward. Given these roles, many researchers have suggested further examination of BPC. However, there have been few attempts to do this, and even fewer attempts to establish and support BPC. These attempts have led to a few models/frameworks of crowdsourcing processes, but these models focus primarily on technical features of crowdsourcing systems rather than the business processes orchestrating on these systems. Furthermore, most of the proposed models so far are inconclusive and thus more empirical research is needed (Amrollahi, 2015; Hetmank, 2013). Thus, what is largely missing in the literature is an informed way to establish BPC, from conceptualising, to modelling, and to empirically supporting BPC establishment.

When initially conceptualising BPC, we note that an antecedent must exist to enable the BPC concept. That is, there exists repeatable building blocks of crowdsourcing processes, which provides the process designers basic elements for creating real life instances of the crowdsourcing process. From the preceding review, we have observed several processes, activities, and components that have been repeatedly discussed and thus can possibly be synthesised into the repeatable building blocks of BPC. The following section explores this possibility, leading us to identify the three highly abstract building blocks.

2.3.3 An Analytical Framework of Business Process Crowdsourcing

Investigating BPC, we now present *an analytical framework* decomposing the concept into its abstract building blocks. The framework mainly draws on the existing literature. Such an analytical framework allows us to channel the related research, and later on, will be used to support our analytical process when we analyse a large number of knowledge sources in the domain to identify repeatable business processes of crowdsourcing.

We start with an abstract view on crowdsourcing activities discussed in previous sections, which, by and large, can be grouped into three high abstract stages: decision to crowdsource, design, and configuration. A crowdsourcing process logically starts with a managerial *decision to crowdsource or not*. This managerial decision considers the appropriateness of crowdsourcing to enhance existing organisational tasks (Afuah & Tucci, 2012; Muhdi et al., 2011; Thuan et al., 2013; Thuan, Antunes, & Johnstone, 2016). After the decision to crowdsource, *design* concerns a set of decisions that have to be made to instantiate a concrete crowdsourcing process. We use the term design to highlight the fact that multiple instantiations are possible and that choice depends significantly on subjective criteria. *Configuration* concerns the materialisation of a design into a concrete system (Kittur, Smus, Khamkar, & Kraut, 2011; Little, Chilton, Goldman, & Miller, 2010). These three stages constitute the analytical framework that presents a logical view of the crowdsourcing process. It is graphically presented in Fig. 2.3.

Each stage of the framework and their main concerns are discussed below.

Decision to Crowdsource
The reviewed literature suggests that a crowdsourcing strategy, like other sourcing strategies, begins with go/no-go decision. This decision is referred to as the decision to crowdsource that considers whether crowdsourcing is appropriate for the organisational tasks (Thuan et al., 2016). Muhdi et al. (2011) and Schenk et al. (2017) position the decision to crowdsource in the first-order position starting the crowdsourcing process. A similar position and purpose of the decision is explicitly stated by other researchers (Djelassi & Decoopman, 2013; Lüttgens et al., 2014; Sandkuhl, Smirnov, & Ponomarev, 2016; Wexler, 2011).

The decision to crowdsource plays a central role in a crowdsourcing strategy, for several reasons. First, it is a strategic decision that directly links to whether an organisation will open or close their boundaries to the crowd (Schenk et al., 2017). Second, it affects the use of organisational resources, at least the resources dedicated to crowdsourcing, because inappropriate decisions are likely to lead to unplanned challenges (Rouse, 2010). Furthermore, as a special kind of project that links to the crowd, a failed crowdsourcing project caused by the decision will influence badly on the organisation's reputation (Thuan et al., 2013). Finally, with

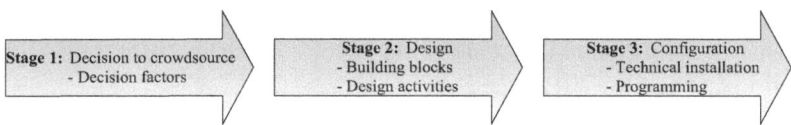

Fig. 2.3 Three-stage analytical framework

its first-order position, the decision to crowdsource cannot be changed, and thus it greatly influences the remaining stages of the entire crowdsourcing processes (Muhdi et al., 2011).

With the first-order position of crowdsourcing, the decision to crowdsource has already received much attention from researchers, focusing on factors driving this decision. Some earlier studies, maybe for simplification, take into account only one factor to make the decision to crowdsource or not. For instance, Ranade and Varshney (2012) addressed the decision "to crowdsource or not to crowdsource" (p. 1) by mainly relying on the factor of task nature. Naroditskiy et al. (2013) examined "the trade-off between the potential for increased productivity with the possibility of being set back by malicious behaviour" (p. 1). However, more recent studies examine a combination of diverse factors in this decision, including benefits and risks (Lu, Hirschheim, & Schwarz, 2015; Muhdi et al., 2011) and organisational structures that founds crowdsourcing operations (Djelassi & Decoopman, 2013). Consequently, the decision to crowdsource is not simple yet complex, where multiple contingency factors should be considered (Afuah & Tucci, 2012; Thuan et al., 2016; Zhao & Zhu, 2014).

Design

After organisations decide to crowdsource, they need to transfer this decision to concrete designs. Design is defined, according to Hevner and Chatterjee (2010), as a plan for structuring elements in order to best accomplish a particular purpose. Adopting this definition, the design stage should plan and structure activities of a crowdsourcing process. It is here the BPC view should maximise its benefits. In other words, this stage should identify both the abstract building blocks, and the detailed design activities and related information structures of the crowdsourcing process.

The literature has shown several possible building blocks of BPC and their detailed design activities. For example, Sect. 2.3.2 has reported a few building blocks proposed by Pedersen et al. (2013) and Hetmank (2013). Another example is the list of design building blocks and activities suggested by Kittur et al. (2013), who research crowdsourcing on complex, large-scale tasks. This list includes twelve abstract activities, including workflow design, task assignment, hierarchy, real-time response, collaboration, quality control, crowds guiding artificial intelligence, artificial intelligence guiding crowds, platforms, task design, reputation, and motivation. Other design building blocks and their detailed activities can also be found in the literature (Ågerfalk et al., 2015; Amrollahi, 2015; Stol & Fitzgerald, 2014; Zogaj et al., 2014).

Given the existence of different building blocks and their detailed design activities, we note here three important points. First, the differences, again, confirms the ad hoc nature of the domain, and thus suggest a more comprehensive integrated approach to synthesise these building blocks. Second, these studies have highlighted the need to design the different building blocks and activities of the crowdsourcing process. That is, to establish a crowdsourcing process, several activities of the crowdsourcing process need to be designed and structured, which suggests the role of process design (Stage 2 of the framework). Finally, despite the differences of the proposed building blocks and activities, we can identify some repeatable activities, such as crowd management (Kittur et al., 2013; Pedersen et al., 2013), how to motivate the crowd (LaToza & Hoek, 2016; Naderi, 2018), and quality control (Amrollahi, 2015; Kittur et al., 2013). Consequently, it is possible and necessary to reconcile the differences and suggest common building blocks of how to design the crowdsourcing process.

Configuration
The configuration stage transforms a crowdsourcing design into a concrete implemented system. In the crowdsourcing context, configuration can refer to either technical decisions to set up crowdsourcing components on existing platforms (Gonnokami, Morishima, & Kitagawa, 2013; Hosseini, Phalp, Taylor, & Ali, 2015a; Kittur et al., 2011; Little et al., 2010), or in-depth technical software development to build a crowdsourcing platform, such as algorithms, protocols, and database structures (Schall, 2012). Although this stage can be considered from both views, the chosen business process perspective limits our concern within the process configuration on an existing platform. This is also supported by the availability of several crowdsourcing platforms (Hirth et al., 2011) and programming toolkits that eases the configuration (Kittur et al., 2011; Kucherbaev et al., 2013; Little et al., 2010; Tranquillini et al., 2015).

Overall, we have synthesised the analytical framework initially conceptualising BPC. The framework structures three high-level stages of BPC: decision to crowdsource, design, and configuration, which will be deconstructed into the main building blocks and activities to thoroughly conceive the BPC concept.

2.3.4 Discussion

The review assessed the literature on crowdsourcing processes and business process crowdsourcing. It identified the two major research streams of crowdsourcing processes: high and low levels of granularity. It finds that some studies research crowdsourcing processes as a whole without its parts, while a large number of studies investigate specific parts of crowdsourcing processes without the whole.

The different levels of granularity have hindered us to have a completed picture of crowdsourcing processes. Further, major reviews of crowdsourcing literature show that the domain is characterised by a large number of scattered, varied and sometimes conflicting knowledge sources. Addressing this challenge requires an integrated view, which has led us to introducing BPC.

Reviewing what little has been published on BPC highlights three important points. First, the review introduces the concept of BPC that brings a business process lens to study crowdsourcing, which enables us to establish crowdsourcing as an organisational business process. Second, the review discusses the roles of BPC. It shows that BPC can resolve the ad hoc challenge and provide structures for the domain. Finally, it finds a few models and frameworks contributing to understand crowdsourcing processes, but not comprehending BPC. Together, these points suggest that BPC is an emerging yet important phenomenon that needs to be conceptualised, modelled, and applied to crowdsourcing practices.

Despite of its early state, BPC is promising to move crowdsourcing from ad hoc processes toward mature repeatable processes. That is, BPC provides a template of repeatable building blocks that organisations can use to instantiate real-life crowdsourcing processes. From the preceding review, we observe that some building blocks that have been repeatedly discussed. Moving this observation forward, we initially synthesise three abstract stages of BPC repeatedly suggested in the crowdsourcing process literature. These stages allow us to channel the related literature in the next chapters to obtain increased insight and thoroughly conceiving BPC.

We note that from the current early state of BPC, this book will engage in conceptualising, modelling, and supporting business process crowdsourcing. The resulting engagement, presented in the remaining chapters of this book, contributes to move the domain to a more mature state, which will be further discussed in Sect. 7.2.4.

2.4   Chapter Summary

This chapter has provided a narrative review to assess the state of the art that driving the book to study BPC. One main drive is that the crowdsourcing domain is emerging, characterised by unstructured knowledge sources and the lack of a strong knowledge base. There appears to be evidence for this in the literature reviewed in the previous sections.

The review covered three major strands. The first strand examined the conceptualisation of crowdsourcing. It shows three main pillars behind the concept, followed by a discussion in order to compare and contrast crowdsourcing with other related concepts. They draw a boundary around the crowdsourcing concept and show that crowdsourcing is a distinctive concept per se. Then the short history of crowdsourcing definitions was discussed to show that the concept continues to evolve. Together, the distinctive concept suggests that crowdsourcing must be developed independently, while the evolution of the concept's definitions and its short history indicate the emerging nature of crowdsourcing.

The second strand reviewed basic classifications in the crowdsourcing domain. This shows that research into classifications cover many particular topics: applications, tasks, types of crowd members and platforms, but not yet cover the synthesis and coordination among them. It is the ad hoc nature of the domain, where classifications are solely relevant to particular crowdsourcing elements or contexts. Further, these classifications have not yet been synthesised and linked in a comprehensive integrated structure, and thus there is still a need for a solid knowledge base that structures the domain.

Finally, the last strand has painted an overall picture of business process crowdsourcing. It shows that BPC is still in an early state with a small amount of related literature, which needs to be further conceptualised, structured, and supported. At the same time, the review shows that BPC is important to establish repeatable crowdsourcing processes, and thus possibly moves the domain toward more mature state. To contribute to this movement, the review has developed an analytical framework presenting the three abstract stages of BPC: decision to crowdsource, design, and configuration. The framework abstractly conceptualises BPC, and will be used to guide our data collection and analysis for further conceptualising and structuring BPC.

The following chapter discusses the main building blocks of BPC.

Chapter 3
Business Process Crowdsourcing: Building Blocks

Specifying building blocks is like giving a process designer a box of Lego stones. He can play around with these stones and create completely new processes, limited only be his imagination and the pieces of stones supplied.

—Adapted from Osterwalder (2004)

This chapter analyses existing knowledge sources for synthesising the main building blocks of BPC. The synthesis is based on what we name the 'wisdom of the researchers' where a collection of researchers is wiser than single experts, similar to the wisdom of the crowd (Surowiecki, 2004). That is, the synthesis focuses on the building blocks that have been suggested by multiple researchers.

For this purpose, we adopt *a scoping literature review* as the main technique of this activity. A scoping review enables a comprehensive view on a particular topic (Paré et al., 2015), and thus is highly suitable for the emerging nature of BPC. More precisely, the 'scoping' review refers to a comprehensive sample strategy, which covers the breadth of knowledge sources existing in the domain. Further, scoping review is explicit in terms of how the search, selection, and data extraction are conducted. This increases the level of transparency and rigour of the research. We note that parts of the scoping literature review have been presented in our conference paper by Thuan et al. (2014).

3.1 Scoping Knowledge Sources

To begin the review, this research established a systematic process to ensure the rigour of the review results. We based the review process on the recommendations of how to conduct a good IS literature review, and especially, a good scoping review (Okoli, 2015; Paré et al., 2015). Following Okoli's (2015) recommendations, we adopted the five steps, including selecting sources, filtering sources, classifying sources, data extraction, and data synthesis. Figure 3.1 summarises the five steps of the scoping review, which are specified below.

© Springer International Publishing AG, part of Springer Nature 2019
N. H. Thuan, *Business Process Crowdsourcing*, Progress in IS,
https://doi.org/10.1007/978-3-319-91391-9_3

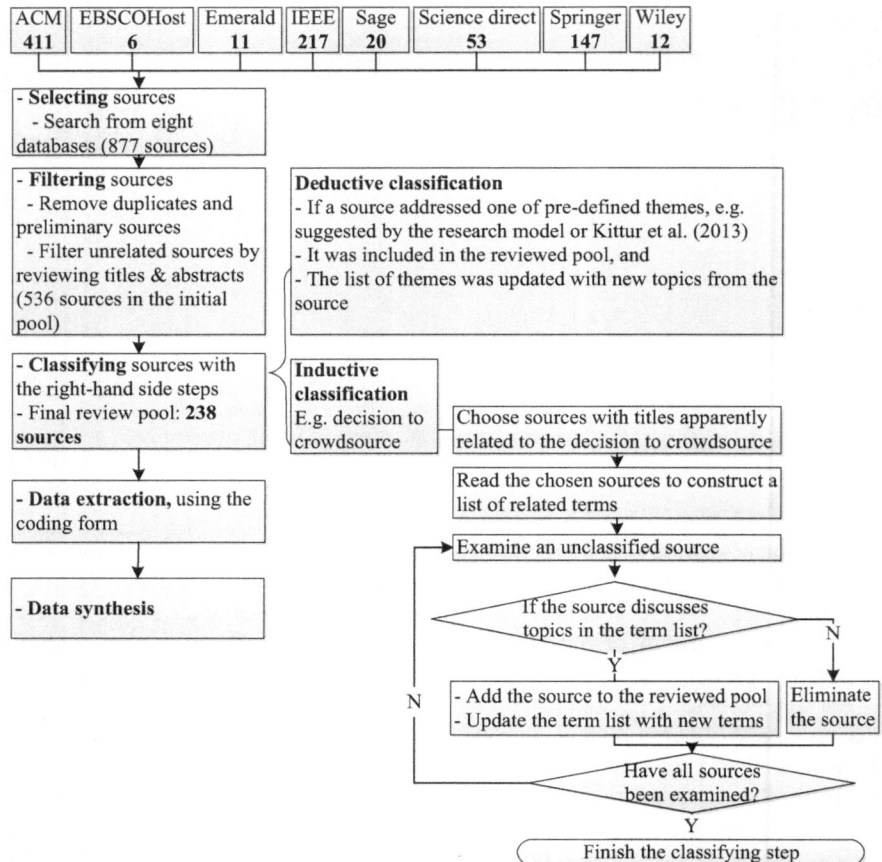

Fig. 3.1 Steps and summary of results of the scoping knowledge sources

3.1.1   Scoping Review Process

Selecting Sources

This initial step searched for the relevant sources about crowdsourcing. Following the scoping approach that highlights the comprehensiveness, the search was opened to multiple knowledge databases. More precisely, it relied on eight popular bibliographic databases: Association for Computing Machinery (ACM), Business Source Elite (EBSCOHost), Emerald Insight (Emerald), IEEE, Sage, Science Direct, Springer Link, and Wiley. In 2013, we searched for papers using the following keywords 'crowdsourcing', 'crowdsource', 'crowdsourced', 'crowdsourcer', and 'crowdsources' (the keyword 'crowdsourc*' was used to replace all the aforementioned keywords in certain allowable databases like Emerald, IEEE, and Sage). This choice of keywords was based on the perception that they are representative and have been popularly used by other reviewers in the domain (Estellés-Arolas & González-Ladrón-de-Guevara, 2012; Hossain et al., 2015). As a result, we identified 877 knowledge sources, consisting of 667 conference papers and 210 journal articles. The search results are represented in the top parts of Fig. 3.1 and detailed in Table 3.1.

Filtering Sources

Although hundreds of sources were identified by the keyword search, many of them were clearly irrelevant to the subject of the book. Following a screening technique suggested by Okoli (2015), this step filtered out the irrelevant sources using the two following actions. First, we excluded posters, tutorials, extended abstracts, and work in progress, papers which are normally too preliminary to be considered as knowledge sources. In this process, we found 22 duplicates that were stored or indexed by more than one bibliographic databases. They were also removed from the pool. Additionally, we eliminated conference papers that had been extended into journal versions to prevent duplication. Second, we eliminated sources applying crowdsourcing to education, medical research, and games with a purpose because these sources have quite a different focus compared to our organisational view. We also eliminated crowdfunding sources, in which organisations raise capital for investments, and thus are distinct from our BPC definition. This elimination was based on the sources' titles, keywords, and abstracts. Through the filtering steps, the list of sources was sharpened into a pool of 536 sources.

Classifying Sources

After excluding irrelevant sources, this step included sources closely related to the research problem. To check whether a source focuses on BPC and thus keep it in the pool, we analysed the source topics. However, codifying topics was not a straightforward task as there was no complete classification frame specifically relevant to BPC. To address this challenge, we conducted a deductive and an inductive classification. In the deductive approach, we generated a list of pre-defined themes, based on the three stages of the research model (Sect. 2.3.3)

Table 3.1 Crowdsourcing sources on the eight bibliographic databases

Bibliographic databases	Number of conference papers	Number of journal articles	Total number of sources
ACM	408	3	411
EBSCOHost	0	6	6
Emerald	0	11	11
IEEE	170	47	217
Sage	0	20	20
Science Direct	0	53	53
Springer Link	89	58	147
Wiley	0	12	12
Total	667	210	877

and research foci suggested by Kittur et al. (2013), like workflow design, task assignment, task design, and quality control. If a source addressed one of the themes, it was kept in the pool and the list of themes was updated with new topics form the source.

The inductive approach was applied in cases where no classification schema could be found. For instance, there was no schema to classify sources related to the decision whether to crowdsource or not. In these cases, we followed the procedure described on the right-hand side of Fig. 3.1. The procedure is illustrated through the following example. First, we started by scanning the pool to choose sources whose titles apparently related to the decision to crowdsource, i.e. a source entitled 'To crowdsource or not to crowdsource?'. Second, we reviewed the chosen sources for identifying the relevant keywords, terms, and themes related to the crowdsourcing decision. This formed a list of terms, which was iteratively updated. Third, every unclassified source was checked to see whether it related to the term list. If so, the source was kept in the pool and the topics addressed by the source were used to update the term list. Otherwise, the source was eliminated. As a result of this procedure, we ended up adding related sources to the reviewed pool and building a term list for the codification.

Overall, we classified *238 sources* related to BPC. We noted that during the classification process there were many cases where sources would broadly refer to the list of terms but present indirect links to BPC. In these cases, a decision to include the sources rather than exclude them was made in order to keep the scoping review comprehensive. Making such decision was also a part of the 'wisdom of researchers', which suggests including diverse opinions that can latter on be collectively aggregated into stronger positions.

Data Extraction

This step extracted and identified building blocks, decision factors, and activities of BPC from the reviewed sources. For this purpose, we developed a coding form. To test the form, a PhD student was asked to code 20 random sources and the results were compared with the researcher's coding of the same sources. This led to small modifications of the coding form. The form codified four dimensions: general information, topic, findings, and application context. First, the first dimension was general information about the source, e.g. reference, year of publication, and whether it is a conference paper or a journal article, which is typically extracted by other reviews (Okoli & Schabram, 2010). Second, we codified the topics using the three stages of the research model and the term list, which was iteratively updated as described in the above section. Another considered dimension was the research findings, which are necessary to understand the BPC process. A part of this dimension included whether findings can be generalised to other situations or are limited to particular contexts (Mingers, 2003). Finally, the last considered dimension codified the practical outcomes of the sources, focussing on useful recommendations about BPC establishment. We also extracted to whom the recommendations were targeted and the crowdsourcing contexts where the outcomes could be applied to.

Data Synthesis

This step aggregated the data extracted by the coding forms. We reviewed the extracted data for building blocks, processes, decision factors, and activities that guided BPC establishment. This was a four-phase procedure. First, we analysed extracted topics and findings for these elements, which were compared and aggregated. Second, we merged the 'conceptually similar' elements. For instance, quality estimation (Baba & Kashima, 2013) and quality control (Allahbakhsh et al., 2013) were merged. Furthermore, many elements were linked to each other, e.g. expert evaluation is a technique to ensure quality control (Allahbakhsh et al., 2013). To rationalise the relationships among them, we mapped some sub-elements into more generic ones. Finally, we synthesised the sources' recommendations that were related to particular elements. As a result, elements extracted from individual sources were synthesised and transformed into thematic elements related to BPC establishment. They are discussed in the next sections.

3.2 Findings

This section reports results from the scoping review. As a result of the previous steps, we identified 238 sources related to BPC. The demographic information of these sources shows that 71% of them are conference papers and 29% are journal articles, which is consistent with the significant role of conference publications in IS and computer science (Freyne, Coyle, Smyth, & Cunningham, 2010). The number of publications per year are presented in Fig. 3.2, which shows a steady increase on the number of crowdsourcing studies published since 2008. This reflects the increasing maturity of the crowdsourcing field. This review also confirms the ad hoc nature of the crowdsourcing field as a large part of the reviewed sources (65%) provide findings that can only be generalised to a similar situation (the bottom parts of the columns in Fig. 3.2). Regarding to whom the implications of the reviewed sources are targeted to, the three most popular ones are managers, process designers, and programmers, which are essentially aligned to the three-stage model discussed in Sect. 2.3.3.

We now report the results of the scoping review in more detail. Considering the purpose of this chapter, we analysed the reviewed sources for building blocks of BPC.

3.2.1 Building Blocks of BPC

Our analysis revealed a diversity of building blocks, which are abstract elements of BPC. In particular, our review identified more than 20 building blocks and their sub-elements. However, the number of sources supporting each of them was highly different. For instance, 'quality control' was supported by more than 40 sources,

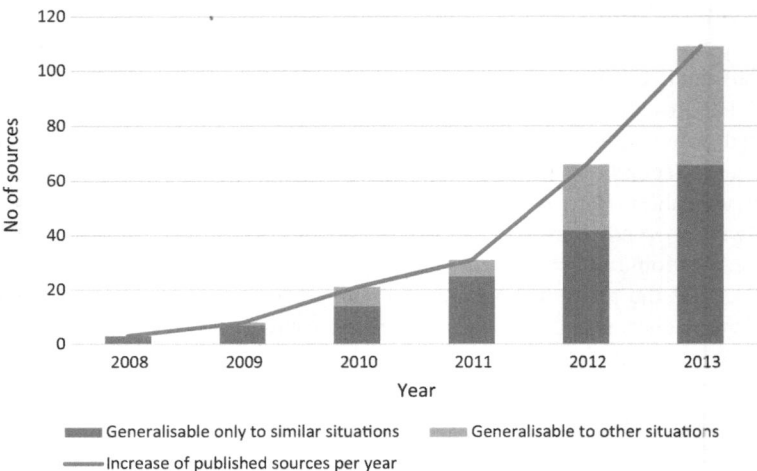

Fig. 3.2 Reviewed sources per year

while some sub-elements like 'guide crowdsourcing with artificial intelligent' were supported by only a few sources. Based on the 'wisdom of researchers' suggesting that aggregating results from groups of researchers outperform individual ones, we refined this list by concentrating only on building blocks supported by more than a certain number of reviewed sources. Choosing this number was quite sensitive. If the number was small, we might include too many building blocks, which unnecessarily increases the complexity of the analysis at this early stage. On the other hand, if the number was large, we might include only a few building blocks and thus might not represent the domain diversity. Given that, we selected a cut-off of 10 sources to balance between representation and complexity. Another reason for choosing this value was that there was a gap between the numbers of supporting sources before (e.g. 16 sources supporting 'circumstance to crowdsource and decision factors') and after the cut-off value (nine and eight sources supporting 'real-time response' and 'benefit & opportunity of crowdsourcing' respectively). As a result, Table 3.2 summarises the main BPC building blocks that are supported by at least ten sources.

From this table, the most popular building block is quality control, which has been suggested by 42 sources. Quality control refers to several techniques ensuring that the "[crowdsourcing] outcome fulfils the requirements of the requester [organisation]" (Allahbakhsh et al., 2013, p. 77). As crowdsourcing workers are voluntary, and thus it is hard for organisations to control their performance, quality control techniques are strongly relevant in a crowdsourcing strategy (Kittur et al., 2013; Zhao & Zhu, 2014). Moreover, incentive mechanisms and crowd management are also popular, being suggested by 37 and 32 sources respectively. To a lesser extent, Table 3.2 also indicates other relevant building blocks of BPC and

Table 3.2 Main building blocks of BPC

Building blocks of BPC	Number of supporting sources (n >= 10)
Quality control	42
Incentive mechanism	37
Crowd management	32
Task design	29
Result aggregation	26
Workflow design	25
Capability and characteristic of crowdsourcing	23
Task assignment	21
Output	17
Circumstance to crowdsource and decision factors	16
Platform	16
Technical configuration	16

their supporting sources, such as task design, result aggregation, workflow design, etc.

Overall, all building blocks identified in Table 3.2 emerge as key elements of BPC. These building blocks indicate repeatable activities within crowdsourcing processes, which backs the BPC concept that considers crowdsourcing as a repeated business process rather than an one-off activity. Further, as the identified building blocks are salient building blocks of BPC, we suggest using them to model and structure BPC, which is the focus of the next chapter. In short, our analysis has identified a set of common building blocks, serving the basic structure of BPC.

3.2.2 Factors Influencing the Decision to Crowdsource

We also identified the important role of the decision to crowdsource in BPC establishment. This important role is partly empirical, given the building block 'circumstance to crowdsource and decision factors' in Table 3.2, and partly theoretical, based on its starting position in the BPC process (discussed in Sect. 2.3.3).

Given the important role, we further analysed factors and sub-factors influencing the decision to crowdsource. The analysis followed the aforementioned review procedure, with two extensions. First, to keep the research up-to-date, we conducted forward searches based on the pool of sources. More precisely, we used the 'cited by' function in Google Scholar to identify the recent publications that cited the sources. The results from these searches increased the number of relevant sources on the decision to crowdsource to *50*. Second, the coding process was slightly modified for identifying directions of influence on the decision to crowdsource. We added quotes on the potential factors and marked '+' for factors that positively influence the decision and '−' for the ones that negatively influence the decision, similar to the method used by Smith et al. (2008). The analysis results, part of which have been presented in Thuan et al. (2016), are now reported in more detail.

Table 3.3 highlights the set of factors that influence the decision to crowdsource. In particular, we found nine main factors, which were then decomposed into sixteen sub-factors or properties. We show how much the knowledge sources support them by presenting the number of supporting and non-supporting sources (the last two columns). We note that the number of supporting sources on a generic factor may be different with the sum of the corresponding references in its sub-factors. This is because in some cases, a source may concern several sub-factors and thus is coded multiple times, while in other cases, some sources study a generic factor as a whole without concerning its sub-factors.

The results from Table 3.3 indicate that 'task' is the most salient factor influencing the decision to crowdsource. 60% of the sources suggest this factor, sometimes under different names such as challenges (Seltzer & Mahmoudi, 2013), problems (Brabham, 2008a; Muhdi et al., 2011), and crowd work (Kittur et al., 2013). This salience is because the task factor is where the substantive decision starts from. It is the first-order question that has to be answered when crowd-sourcing (Malone et al., 2010). This factor is also important as it determines several aspects of a crowdsourcing strategy, including the targeted crowd that has the ability to perform the task, the chosen platform for publishing the type of tasks, and the internal experts supporting crowdsourcing activities. Table 3.3 also presents seven sub-factors of tasks. Four of them positively influence the decision to crowdsource: whether tasks are easy to delineate (10 sources), to partition (8 sources), to integrate with existing business processes (7 sources), and to be done through the Internet (5 sources). Three other sub-factors negatively influence the decision: whether the task includes confidential information, needs high interaction, or can be automated.

Besides the task, two factors that are most addressed by the reviewed sources are the availability of the crowd and risk. The crowd, which comprises who will perform a task, was found in 38% of the reviewed sources. These results are not surprising because the crowd is one of the three key underpinnings behind the crowdsourcing concept, as shown in Chap. 2. Out of 50, 14 sources suggest the risk factor, which has a negative impact on the decision to crowdsource (i.e. more risk means less opportunity to crowdsource). To a lesser extent, other factors like infrastructure availability, availability of crowdsourcing experts to manage tasks, budget, internal human resources, and internal commitment also seem to influence the decision to crowdsource. Lastly, the level of organisations' technology adoption is the least addressed factor.

In summary, the review allowed us to systematically identify a set of factors that influence the decision to crowdsource. Using the identified factors, we can evaluate whether BPC is a suitable approach for a particular organisational context. Yet the relationships, similarity, and disparity among these factors still need to be examined and structured, which will be examine in the next chapter.

Table 3.3 Factors that influence the decision to crowdsource

Decision factors	Factor's properties/ sub-factors	Number of supporting sources	Number of non-supporting sources
Task		30	1
	Ease of delineation	10	
	Partitionable	8	
	Ease of integration with existing business processes	7	
	Done through the Internet	5	
	Confidential information (−)	3	1
	High interaction or requiring training (−)	2	
	Hard to be automated	1	
Availability of the crowd to perform the task		19	
	Number of members	9	
	Diversity	6	
	Knowledge	5	
	Internet access	3	
Risks (−)		14	1
	Low quality results (−)	8	
	Loss of intellectual property (−)	4	1
Infrastructure		12	
	Availability of crowdsourcing platform	10	
Expertise to manage the crowdsourcing activity		6	
Small budget		4	4
Lack of internal human resources to accomplish the task		3	
	Number of employees	3	
	Knowledge	2	
Lack of internal commitment (−)		3	
Slow in technology adoption (−)		1	

3.3 Summary and Discussion

This research conducted a scoping review of domain knowledge sources through a systematic process. The process retrieved 877 sources from eight bibliographic databases and finally considered 238 sources relevant to BPC. An overview on the reviewed sources confirmed the ad hoc nature of the BPC domain, which has supported the motivation of the book to study BPC. Analysing the sources in detail, the results revealed and synthesised the major building blocks of BPC. Of them, there were twelve most salient BPC building blocks supported by at least ten reviewed sources (Table 3.2). The analysis also identified factors influencing the decision to crowdsource. It revealed nine factors and sixteen sub-factors that should be considered in the crowdsourcing decision (Table 3.3). The identified building blocks, decision factors, and synthesised knowledge provide raw materials for the next research stages.

Overall, the scoping review offers accumulated knowledge of what the literature has reported in the domain. It has confirmed that there are repeatable processes of crowdsourcing strategies, through the identification of building blocks repeatedly suggested by the knowledge sources. The repeatable processes are the important antecedent of BPC and properly constitute business processes of crowdsourcing. Regarding the nature of the review, since the review process was arranged systematically and presented explicitly, it is possible for the review process and its results to be reproduced. This increases rigour of the review process and adds confidence to the review results. All in all, a combination of knowledge accumulation and systematic-ness constitutes the value of the scoping knowledge sources.

Chapter 4
Business Process Crowdsourcing: Model and Case Study

In emerging areas, the author's contribution would arise from the fresh theoretical foundations proposed in developing a conceptual model.
—Adapted from Webster and Watson (2002).

Using the raw materials extracted from the knowledge sources, this chapter articulated and built a conceptual model supporting the establishment of crowdsourcing as an organisational business process. Such a conceptual model had important roles in this research. The model, which articulated the raw materials into organised BPC information, provided an abstract and holistic view on the BPC domain (Cross, 1982). With its articulation, the model also underpinned the conceptualisation of BPC, and thus provided a means to explore the field. This role has been suggested by Hevner et al. (2004) that design science research may start with "simplified conceptualizations and representations of problems" (p. 85). The role of the conceptual model should also be seen as a research outcome, where a conceptual model constitutes an IS artefact per se (Hevner et al., 2004).

As the built model served as an IS artefact, it should be rigorously evaluated. The current chapter evaluated the model using a case study approach. More precisely, this evaluation considered the model in two crowdsourcing projects, which confirmed the adequateness and utility of the model. When considering this evaluation in the research process, the case study provided empirical evaluation of the model, which complemented the previous research efforts to conceptualise BPC. We note that this chapter is based on the journal publication by Thuan et al. (2017) with further details.

4.1 A Process Model for BPC Establishment

To build the conceptual model, we followed guidance from Webster and Watson (2002) and Jabareen (2009) for conceptualising models from extant literature. These authors suggest that a conceptual model can be built and generalised based

© Springer International Publishing AG, part of Springer Nature 2019   49
N. H. Thuan, *Business Process Crowdsourcing*, Progress in IS,
https://doi.org/10.1007/978-3-319-91391-9_4

on a literature review. In particular, Webster and Watson (2002) suggest analysing the related literature for main concepts and processes, which are main materials for model construction. Agreeing with this suggestion, Jabareen (2009) further recommends viewing a conceptual model as not only a simple set of concepts, but rather as an organised structure where each concept plays an integral role. Following these suggestions, the current research used the key building blocks drawn from the scoping review, and structured them in a meaningful way. Since these building blocks were repeatable processes of crowdsourcing, this structure led us to construct *a process model of BPC*.

We structured the original BPC building blocks (Table 3.2) to construct the process model of BPC. However, structuring these building blocks was not a straightforward task, since they covered very different concerns. Addressing this difficulty, the three-stage framework discussed in Sect. 2.3.3 was used as a starting point for the structuring process. We tried to allocate each building block into one of the three stages: decision to crowdsource, design, and configuration. The allocations on the decision to crowdsource and configuration were transparent, because they exhibited strong conceptual links. For instance, building blocks such as 'circumstance to crowdsource and decision factors' and 'characteristic of crowdsourcing' were logically linked to the decision to crowdsource. Similarly, 'technical configuration' was also clearly linked to the configuration activity.

However, allocations of building blocks to the design activity were more difficult since the links extracted from the reviewed sources were more diffuse. To help logically organise the building blocks, we classified these building blocks into plan-time and operation-time categories according to when they are processed. The 'task design' and 'workflow design' were related with the plan-time category, as they should be done before the tasks are sent to the crowd. The remaining building blocks, including 'crowd management', 'quality control' and 'incentive mechanism' included activities that are operationalised while the crowd performs tasks. In particular, crowd management includes profiling the crowd; quality control includes identifying cheating behaviours; and incentive mechanism includes dynamic pricing, all of which process information while the crowd performs tasks. As a result, this structuring organisation led to the process model shown in Fig. 4.1.

We now describe the process model in more detail. As seen in Fig. 4.1, the model adopts the input-process-output (Pedersen et al., 2013) and stage-gate configurations (Cooper, 2008) that are typical of process models. It consists of seven components structured into three stages, which are described as follows.

Decision to crowdsource. The crowdsourcing process is triggered by an opportunity to crowdsource a piece of work, which starts a *decision to crowdsource*. This component initially conceptualises the crowdsourcing strategy in order to "decide whether the crowdsourcing approach is appropriate to solve their internal problem/problems [tasks]" (Muhdi et al., 2011, p. 322). It is a logical antecedent to any crowdsourcing project, aligning to a 'make or buy' decision in outsourcing projects. By making it explicit in the model, we signal that the decision to crowdsource should be founded on a logical assessment of the crowdsourcing context adequacy.

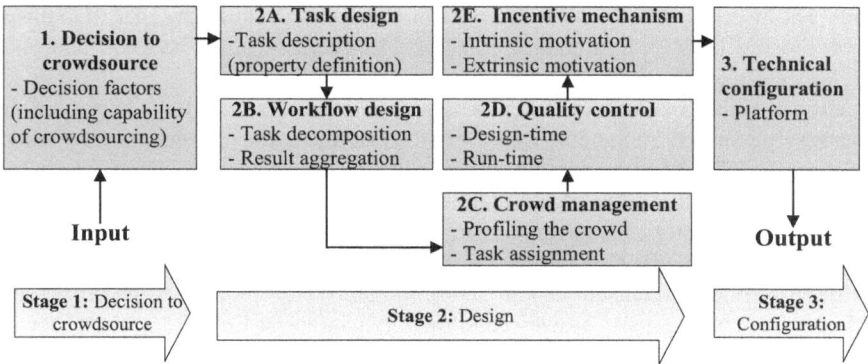

Fig. 4.1 A process model for BPC establishment

To make a logical decision to crowdsource or not, organisations need to evaluate several decisional factors. Table 3.3 has already identified several factors influencing the decision to crowdsource. However, we note that many factors in Table 3.3 may link to each other, which needs to be further arranged. Given that, we decided to structure these factors into a decision framework in order to support managers making informed decisions when they come to crowdsourcing. Yet, to keep the flow of the current section focusing on the process model and due to the important role of the decision to crowdsource, we present this framework separately in the next section.

Design. After the decision to crowdsource has been made, this stage covers a set of design activities necessary to operationalise the decision. It includes five components: task design, workflow design, crowd management, quality control, and incentive mechanism. *Task design* aims at transforming the conceptual ideas about the crowdsourcing tasks into a concrete task description (Model component 2A). Most of the reviewed sources recommend clearly defining the tasks that are crowdsourced (Malone et al., 2010; Rosen, 2011). The aim of this component is to designate a complete task description that can be given to the potential crowd members who may perform the tasks. To define these tasks, the properties suggested by Zheng et al. (2011) and Tokarchuk et al. (2012), like significance, autonomy, etc., should be taken into account.

The next component concerns *workflow design*. This involves task decomposition and result aggregation (Model component 2B). The former decomposes the list of tasks into smaller tasks, which can often be performed with massive parallelism. This decomposition increases the potential number of workers interested in participating in the open call (Afuah & Tucci, 2012; Kulkarni, Can, & Hartmann, 2012). A counterpart of decomposition is result aggregation, which concerns the definition of how the outputs from the smaller tasks will be put together so that the objectives of the overall task may be fulfilled (Geiger et al., 2011). Result aggregation is closely linked to task decomposition as they are two sides of the same coin. Kittur et al. (2013) explain this relationship as a workflow that "facilitates decomposing tasks into subtasks, managing the dependencies between subtasks, and assembling the results" (p. 5).

Crowd management is a design component that refers to how organisations manage the crowd members in order to accomplish the defined tasks (Model component 2C). The reviewed sources suggest two sub-components of crowd management: profiling the crowd and assigning tasks. First, organisations analyse the required capacity of crowd members for performing a task (Allahbakhsh et al., 2012; Kittur et al., 2013), and use this evaluation to build member profiles. Based on these profiles, organisations can determine an overall picture of the crowd and may impose constraints to crowd recruitment (Chandler & Kapelner, 2013; Stewart et al., 2010). Second, based on the crowd profiles, task assignment can be executed. That is, tasks can be assigned to crowd members who have appropriate profiles. Examples of existing task assignment mechanisms include the auction-based mechanism (Satzger et al., 2011) and the scheduled mechanism (Khazankin, Satzger, & Dustdar, 2012b).

According to Table 3.2, *quality control* should be regarded as the most critical model component (Model component 2D). One distinctive characteristic of crowdsourcing is that tasks may be performed by crowd members with very different backgrounds, skills and expertise (Hirth, Hoßfeld, & Tran-Gia, 2012). This sometimes leads to a number of low-quality contributions. Thus quality control mechanisms are critical to ensure the outputs meet the organisation's quality goals (Allahbakhsh et al., 2013; Ipeirotis et al., 2010). By and large, quality control mechanisms can be grouped into design-time and run-time mechanisms (Allahbakhsh et al., 2013). At design-time, organisations can design tasks and workflows in a robust way, to increase the chances of receiving high-quality contributions. For instance, Eickhoff and De Vries (2013) recommend that defining tasks in an unambiguous and abstract thinking way can increase quality contributions. At run-time, organisations can consider several active quality control mechanisms like expert reviews, peer reviews, gold standards, output agreements, and even peer assessments with majority voting (Allahbakhsh et al., 2013).

Crowdsourcing relies on members of the crowd voluntarily performing tasks. Thus, organisations need *incentive mechanisms* to attract and engage these voluntary members in their open calls (Model component 2E). The reviewed sources suggest that incentive mechanisms should be developed based on two main types of motivation: intrinsic and extrinsic. For extrinsic motivation, most of the investigated sources have examined the adoption of financial incentives (Kaufmann et al., 2011; Mason & Watts, 2009). Regarding intrinsic motivation, a variety of factors have been suggested by the extant literature, such as fun (Doan et al., 2011), meaningful tasks (Chandler & Kapelner, 2013), and love of the community (Kaufmann et al., 2011).

Configuration. The final component considers how to *configure* a crowdsourcing process for instantiation in computational systems. Since this activity mainly concerns an in-depth technical view, for instance, adopting specific architectures, frameworks, and proprietary or open computational platforms, the business perspective adopted by this study limits our considerations regarding this component. Besides, since several crowdsourcing platforms are readily available, we expect this component to be significantly constrained by the service providers. Furthermore, we note the extant literature has already proposed several tools supporting the configuration process. That is, we expect that in the near future, given a designed crowdsourcing process, tools may be able to automatically translate such designs into process instantiation capable of running on specific crowdsourcing platforms. Examples of such translation tools include Turkit (Little et al., 2010), Crowdforge (Kittur et al., 2011), and BPMN4Crowd (Tranquillini et al., 2015). Given that, we regard the main output of this component as a configuration file necessary for implementing the crowdsourcing process, but we do not further research the low-level details that already examined by the translation tools.

4.2   A Framework Supporting the Decision to Crowdsource

The decision to crowdsource plays an important role in the crowdsourcing process. This role has been highlighted in the process model positioning the decision to crowdsource as the first component starting BPC (Fig. 4.1). A similar role has been supported by several researchers (Lu et al., 2015; Lüttgens et al., 2014; Muhdi et al., 2011). Given the importance, researchers have proposed several factors influencing the decision to crowdsource, which have already been identified and summarised in Table 3.3.

In this section, we used the identified factors to build an analytical framework for supporting the decision to crowdsource. To this end, the 'wisdom of researchers' was applied to Table 3.3, leading to the elimination of factors suggested by only one reviewed source and focusing on factors suggested by multiple sources. We then structured the remaining factors in a meaningful manageable way. Specifically, we adapted the multi-layer approach proposed by Vicente (1999), which highlights the multiple concerns that need to be understood in the decision. Consequently, we classified the decision factors in four layers, including the task, people, management, and environment. These layers are depicted as a decision framework in Fig. 4.2. The framework has presented in Thuan et al. (2016) and is further explained below.

Task Properties. According to Table 3.3, the reviewed sources suggest tasks as a key factor in the decision to crowdsource (Kazman & Chen, 2009; Rouse, 2010; Zhao & Zhu, 2014). From these sources, using the crowd may be good for certain tasks, but not for all kinds of tasks. Consequently, it is critical to examine task characteristics for evaluation whether an organisational task is suitable to be crowdsourced or not (Muntés-Mulero et al., 2013). This key role leads us to position this factor in the core layer of the framework. In this layer, we define six task properties.

The first property is whether a task can be performed or delivered online, i.e. its inputs/outputs can be delivered and collected through the Internet. Most of the reviewed sources consistently suggest that crowdsourcing should only be used for Internet activities (Brabham, 2008a; Doan et al., 2011; Muntés-Mulero et al., 2013). Some researchers go further adding this property to the definition of crowdsourcing, which turns this factor into one of the key underpinnings of crowdsourcing activities (Sect. 2.1.1).

The second property concerns the integration between crowdsourcing and the existing organisational business processes. This integration tightens and coordinates the external tasks and internal business processes (Tranquillini et al., 2015), which is strongly aligned with the BPC perspective of the book. Furthermore, the important role of this factor is supported by several reviewed sources, which suggest examining not only individualised crowdsourcing tasks but the whole business process (Kittur et al., 2013; Sakamoto, Tanaka, Yu, & Nickerson, 2011). The importance of this factor has increased recently due to the increasing adoption of crowdsourcing for complex organisational processes, including product development processes (Djelassi & Decoopman, 2013), industrial problems (Muntés-Mulero et al., 2013), and software development processes (Mao et al., 2017; Stol et al., 2017).

Interaction is the third considered property, which focuses on the ties between the organisation and the crowd members during crowdsourcing activities. Overall, a

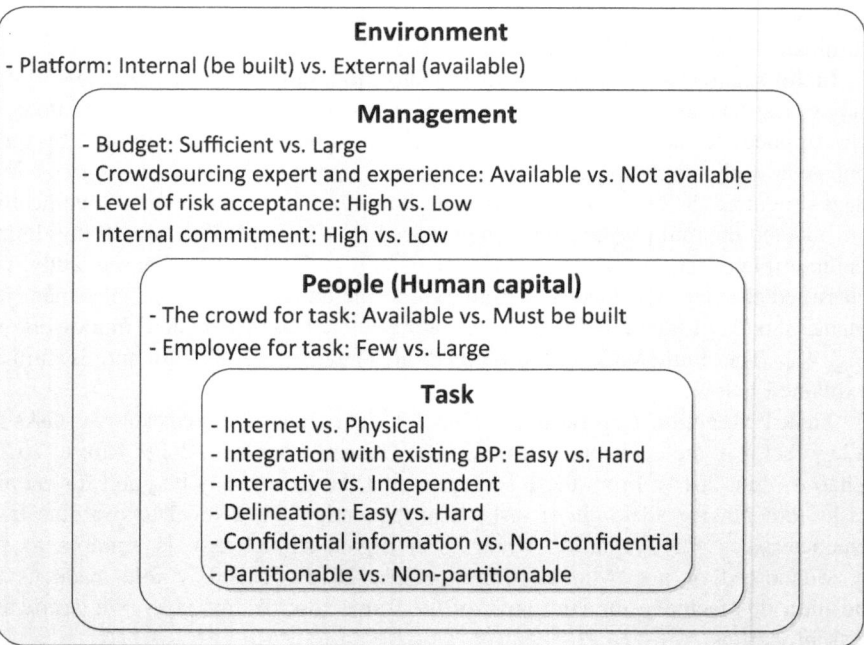

Fig. 4.2 A framework that supports the decision to crowdsource

decision to crowdsource seems unsuitable for interactive tasks that require frequent exchanges between the organisation and the crowd, or between members of the crowd (Burger-Helmchen & Pénin, 2010). The reason is that it is quite hard to promote interaction when the crowd members are anonymous agents (Afuah & Tucci, 2012). Similarly, Muntés-Mulero et al. (2013) also suggest avoiding crowdsourcing if complex training is required to fulfil a task. As a result, independent tasks that do not require a lot of interaction and training to be accomplished are more compatible to crowdsource.

Ten out of fifty reviewed sources highlight the fourth property, 'ease of delineation', in the decision to crowdsource (Table 3.3), which considers how the task is defined and scoped. Zogaj et al. (2014), Seltzer and Mahmoudi (2013), and Lloret et al. (2012) all suggest the positive influence of this property on the decision to crowdsource. More precisely, organisations should adopt a crowdsourcing strategy when they have well-defined and clearly-scoped tasks. The ease of delineation helps maximise the potential number of workers by increasing the crowd's understanding and so improve their approach to the task (Afuah & Tucci, 2012). It is worth noting that task delineation may have different levels of detail, according to different stages of the crowdsourcing process, from highly abstract in the decision to crowdsource to more specific in the design and configuration.

The fifth property is whether or not tasks include confidential information, which could result in privacy and security issues. Since crowdsourcing tasks are usually sent to anonymous members of the crowd, Muntés-Mulero et al. (2013) argue that tasks with confidential information are not suitable for crowdsourcing. In a similar vein, Burger-Helmchen and Pénin (2010) suggest that the decision to crowdsource should only be made if intellectual property rights can be clearly defined. Although agreeing with the suggestion, other researchers believe that additional efforts may deal with and mitigate the problem of sensitive information. Lu et al. (2015) and Feller et al. (2012) suggest decomposing tasks into a large number of smaller tasks to conceal the overall picture, which decreases the likelihood of privacy breaches and claims regarding intellectual property.

The sixth and final property is the ease with which a task can be partitioned into smaller pieces of work. The influence of this property on the decision to crowdsource is suggested by several reviewed sources. Malone et al. (2010), when discussing the collective intelligence of the crowd, point out that a crowdsourcing strategy is more adequate for tasks that can be partitioned. Similarly, Afuah and Tucci (2012), regarding problem-solving tasks, hypothesise that this property positively influences probability of choosing a crowdsourcing strategy. Furthermore, this property indirectly affects the decision to crowdsource through strengthening the other aforementioned properties. Partitionable tasks are expected to be easier to delineate (Feller et al., 2012) and to protect sensitive information (Lu et al., 2015), each of which positively influences the decision to crowdsource.

People. When making the decision to crowdsource, an organisation should consider the role of human capital playing in the crowdsourcing process, in terms of the crowd members and internal human resources (Afuah & Tucci, 2012). The availability of the crowd members to perform tasks is the key factor deciding the choice of crowdsourcing as tasks in the crowdsourcing strategy are processed by the crowd members. In general, Djelassi and Decoopman (2013) and Doan et al. (2011) suggest that the high availability of members increases the possibility of adopting a crowdsourcing strategy. Afuah and Tucci (2012), examining crowd-sourcing contests, identify a similar positive influence.

The availability of the crowd should be further considered through four sub-factors: the number of members in the crowd, Internet access, knowledge, and diversity. According to Table 3.3, the number of members and their ability to access the Internet are two determinants for crowd availability. Both Malone et al. (2010) and Marjanovic et al. (2012) indicate that the chance of an organisation choosing to crowdsource increases when there is a large pool of people to procure for the task. The requirement of Internet access within the targeted crowd is related to the fact that almost all crowdsourcing tasks are performed through the Internet. Consequently, Internet access influences the number of members available for crowdsourcing tasks (Brabham, 2008a; Saxton et al., 2013), and thus affects the decision whether to crowdsource or not. The other two sub-factors, i.e. knowledge and diversity, also play an important role in the crowd availability. Yet, their roles seem to depend on the nature of the task. For instance, some tasks, like software development (Stol & Fitzgerald, 2014), require a certain type of knowledge from the crowd members, while others, such as solving a generic problem or innovation (Boudreau & Lakhani, 2013), need a crowd with diverse backgrounds. In short, the decision to crowdsource is influenced by "the constant availability of sufficient quantity and quality [knowledge and/or diversity] of online workers" (Corney et al., 2010, p. 244).

The reviewed sources also suggest considering the availability of internal employees when making the decision to crowdsource. If an organisation has too few internal employees in comparison to large human resources required for the task, choosing crowdsourcing to fulfil the human resource gap is suggested (Malone et al., 2010). Lu et al. (2015) go further to explain this decision in terms of both number of employees and their knowledge for tasks. With some tasks, like image tagging and translation, requiring a huge number of human resources that often exceed an organisation's capability, crowdsourcing is a good (if not the only) option. Agreeing with the suggestion, Afuah and Tucci (2012) further considered the internal human resources regarding whether the knowledge meets the require-ments for tasks. Consequently, they recommend using crowdsourcing if "the knowledge required to solve the problem falls outside the focal agent's knowledge neighbourhood" (Afuah & Tucci, 2012, p. 369).

To sum up, the framework suggests that both high availability of the crowd and scarcity of internal employees for the tasks increase the possibility to choose crowdsourcing. When comparing the two factors, the availability of the crowd should receive higher priority. The reason is that the crowd is one key underpinning of crowdsourcing (Sect. 2.1.1), which is again highlighted here by many review sources, i.e. nineteen out of fifty sources in the reviewed pool, compared to three sources suggesting the role of scarce internal employees. Furthermore, though organisations may have enough internal employees for tasks, crowdsourcing is still a good approach that can bring competitive advantages for the organisations, e.g. increasing customer relationship. This can be inferred from many existing crowdsourcing projects promoted by well-resourced organisations, like Westpac bank (Westpac, 2013).

Management. Whether to crowdsource or not is a complex decision, which can influence the success of the whole project. Thus, it has to receive major attention from managers (Djelassi & Decoopman, 2013). From a managerial perspective, Rouse (2010) advises that the decision to crowdsource should only be made after examining costs, coordination, and risks. Recent studies additionally suggest that employees' commitment is another factor influencing the decision to crowdsource (Lüttgens et al., 2014; Simula, 2013). Consequently, the management layer in our framework focuses on four factors: the project budget, the availability of expertise to coordinate the crowdsourcing activity, risks, and internal employees' commitment.

When evaluating whether crowdsourcing is a suitable strategy, it is important to compare its efficiency in realising organisational goals in comparison with other alternatives. As cost saving is a key criterion for measuring efficiency (Muhdi et al., 2011), the budget of a crowdsourcing project influences the decision to crowdsource. Although there is a high agreement on the important role of budget in the decision, the reviewed sources seem to disagree on how this factor influences the decision to crowdsource. As seen via Table 3.3, four sources suggest a low budget, whereas an equal number of sources suggest a reasonable budget before making the crowdsourcing decision. In particular, some sources support that crowdsourcing is a preferred option when a project does not have enough money to hire new employees, or is a small-budget project (Malone et al., 2010). Whereas, others argue that a reasonable budget is required because though the amount of money to pay the crowd may be small, other costs, like coordination and transaction costs, may increase (Lu et al., 2015). Although further studies are needed to solve this disagreement, we suggest that the decision to crowdsource should be made based on having sufficient budget. That is, the budget is not enough to perform tasks in the traditional way, i.e. internal sources and outsourcing, but is sufficient to cover the crowdsourcing process.

Another considered factor in this layer is whether organisations allocate appropriate expertise and experience to coordinate multiple activities of crowdsourcing. This factor greatly influences the success of crowdsourcing, as stated by Muhdi et al. (2011) that at the beginning of a crowdsourcing project, "a source of experience and expertise in crowdsourcing can be helpful to match company expectations and the realistic possibilities of crowdsourcing" (p. 323). As Rouse (2010) suggests, a lack of coordination can lead to a drain of resources and substantial delays.

By analysing the reviewed sources, we have identified a few risks that should be considered when deciding to crowdsource. According to Table 3.3, the most salient ones are the risks of low quality results (Kannangara & Uguccioni, 2013; Naroditskiy et al., 2013) and loss of intellectual property (Schenk & Guittard, 2011). In crowdsourcing where tasks are performed by voluntary crowd members, organisations have little control over members' behaviour (Zhao & Zhu, 2014), and this could lead to poor contributions to the project. As a result, the risk of low quality results should be considered. Another risk is the loss of intellectual property (Marjanovic et al., 2012), which mainly links to skilled tasks. When relying on the crowd members for these types of tasks, organisational knowledge may have to be transferred to them (Afuah & Tucci, 2012) and after the tasks are accomplished, knowledge related to the task may remain in the crowd. This implies the risk of losing intellectual property. Burger-Helmchen and Pénin (2010) claim that crowdsourcing should only be seen as a viable option if intellectual property can be managed and controlled. We further note that managing intellectual property is not only about hiding sensitive information, as mentioned in the task layer, but can be extended to other mechanisms, such as patents (Burger-Helmchen & Pénin, 2010) and intermediary platforms (Feller et al., 2012). In summary, organisations have more chance of making the decision to crowdsource if they can accept and manage the two aforementioned risks.

The fourth and final factor we consider in this layer is the organisational employees' commitment to crowdsourcing activities, a concern suggested by recent studies (Lüttgens et al., 2014; Simula, 2013). This factor refers to the conflicting interests of employees and managers regarding the crowdsourcing activity, which relates to overcome the issue of the 'not invented here syndrome' (Katz & Allen, 1982). Although only a few articles in crowdsourcing literature consider this factor, we believe it is an important managerial concern because limited organisational employees' commitment "can jeopardise the success of an entire crowdsourcing project" (Muhdi et al., 2011, p. 322). This factor is further important as several tasks in a crowdsourcing project, such as task definition and workflow design, are performed internally by organisational employees and managers (Whitla, 2009; Zhao & Zhu, 2014). As a result, a lack of employees' commitment may decrease the ability to choose crowdsourcing (Lüttgens et al., 2014).

Environment. The primary factor in this layer is the choice over the use of either internal or external crowdsourcing platforms. In terms of cost, using an external platform saves development cost, which makes the decision to crowd-source more competitive. From a resource-based view, Lu et al. (2015) support this argument by clearly specifying that "decisions on the use of online microsourcing [crowdsourcing] will be driven by the ability of online sourcing platforms to provide cheap service solutions, complement current resources, fill a resource gap, and to give access to a large pool of resources" (p. 4). Some other reasons to adopt external platforms include the large and varied pools of members, the speed of launching the crowdsourcing project, and in some cases, protecting intellectual property (Feller et al., 2012; Mason & Suri, 2012; Zogaj et al., 2014).

To sum up, the decision framework developed in this section has two characteristics. First, it structures the factors influencing the decision to crowdsource into the corresponding layers, of task, people, management, and environment, which are not apparent in individual sources of knowledge. Consequently, it can be used as a decision framework per se, supporting managers in their crowdsourcing decisions.

Second, the framework details the first component of the process model (Fig. 4.1), and thus can also be seen as an integrated plugin of the process model.

4.3   Case Studies

After the construction of the process model, we now evaluated the model using case studies. The decision to use case studies was driven by three reasons. First, case studies allowed the model to be evaluated in the practical organisational environments, which is the target application of the model. Another reason came from the complex nature of crowdsourcing. Evaluating a model that captured such a high level of complexity required in-depth and detailed explanations about their components, links and overall structure. The capacity to discuss the model in such detail was a distinctive characteristic of case studies. These reasons were supported by Yin (2013b), who stated that *"for evaluations,* the ability to address the complexity and contextual conditions nevertheless establishes case study methods as a viable alternative among the other methodological choices" (p. 322). The third and final reason was that case studies are considered appropriate for evaluating design science artefacts in complex organisational settings (Peffers, Rothenberger, Tuunanen, & Vaezi, 2012).

4.3.1   Overview of the Approach

To evaluate the model, we had to choose its evaluation metrics. In particular, we considered the two metrics: *adequateness* and *utility* of the model. We defined adequateness as 'the degree to which the components and their arrangement in the model align with the activities done in the studied crowdsourcing project', and utility as 'the usefulness of the model perceived by the crowdsourcing project managers and coordinators'. Using these two metrics, we collected and analysed data from two crowdsourcing projects.

4.3.2   Case Study Design

We followed the guidelines provided by Yin (2013a, 2013b) for designing case study evaluation research, including how to select cases, collect data, analyse data, and validity.

Case Selection

The selection of crowdsourcing projects was based on comparability and access to source material. First, we selected projects with a comparable team size, between 2 and 10 members. This range of team size was sufficiently large to include multiple project roles, which the model aims to support, but not so large as to hold a diversity of settings that overshadow the evaluation purposes. Second, we chose crowdsourcing projects where we had access to project participants and other data sources. As a result, two crowdsourcing projects, Crowd Tagging (CT) and Logo Design Contest (LDC), were selected.

The CT project was part of a bigger plan aiming to uncover the impact of New Zealand predators on biodiversity in urban areas. This plan involved the installation of motion-triggered cameras in 40 locations in New Zealand, which collected more than 65.000 pictures. The CT project aimed at identifying the animals captured in these pictures. Because of the large number of pictures that needed to be analysed, the project launched a website with an open call to help tag the pictures. The project involved a team of four members: project manager, designer, web developer, and consultant. The call went live from June to December 2014. As a result, the project attracted over 300 users. About half of them tagged more than 20 pictures.

The other project, LDC, utilised the crowd for artistic design. A University in the Mekong delta, Vietnam was founded in 2013 from what began as a tertiary education centre. As a part of this transformation process, the University needed a new logo that would represent the spirit of the University. To design the logo, the University adopted a crowdsourcing approach that opened the logo design to designers from both inside and outside the University. It was in this spirit that the LDC project was created. The project started in May 2013 and finished in December 2013, when the winning logo was officially adopted by the University. The project had a leader, who made all project decisions, and a coordinator who instantiated and controlled the contest. The project also involved the University Board, consisting of eight members, who made key strategic decisions about the project planning. When the project was launched, it received 68 logo designs from the crowd. Three of them were selected and declared as the winning solutions: two were awarded for creative prizes and one was awarded for the final winning solution, which is the current logo of the University.

Data Collection

We collected data from multiple sources, both primary and secondary. Secondary sources included press releases, the open calls, meeting reports, and project websites, all of which provided materials necessary to clarify key project activities. The activities and their relationships were further detailed and validated in interviews. Across the two case studies, we conducted three in-depth interviews with project leaders and other participants, both face-to-face and through Skype. Due to the small size of the project teams, these interviewees wore 'many hats' and therefore could provide insights into several perspectives of the crowdsourcing projects. Besides being interviewed about the activities performed in the projects, the interviewees were asked to analyse a printed version of the model presented in Fig. 4.1 and were asked to make a judgment and produce comments about the usefulness of the model. A summary of demographic information about the cases and their data sources is presented in Table 4.1.

Data Analysis

To prepare data for analysis, we first arranged a full description of each case, including details about the project, project team, and project activities. We then used the process model to map the project activities into the model components, while critically analysing the interviewees' comments about the model. More precisely, this empirical analysis included the two following activities.

Adequateness analysis: This analysis followed a pattern matching technique (Yin, 2013a). We looked for major similarities, patterns, and notable differences between the model components and the activities reported for each project. We analysed each project starting from secondary data, which included considerable information about the project activities, followed by the analysis of the interview and supplementary materials. The identified activities were finally mapped in the model for comparing the similarities and differences between them. As a result, the final list of matching patterns (both similarities and differences) was created, allowing us to map the project activities in the model for comparing between them (presented in Figs. 4.3 and 4.4).

Utility analysis: We gathered judgements and comments from the interviewees regarding the perceived utility of the model. During the interviews, we asked evaluation questions, such as 'what do you think about the model components?' and 'what do you think about the sequence of the model components?'. Analysing answers of these questions, we then focus more on identifying patterns of 'usefulness', 'future use' and 'future improvement', rather than 'yes or no' answers as these direct answers are usually biased, which will be discussed in the next section.

Table 4.1 Demographic information about the two crowdsourcing cases

Dimension	Crowd tagging (CT)	Logo design contest (LDC)
Number of project members	3	10
Project duration	6 months	7 months
Project purpose	To tag pictures about animals in New Zealand	To design a logo for the University
Interviews	1	2
Roles of interviewees	• Project leader	• Project leader • Project coordinator
Other data sources	• Press and media • Website, tutorial • Internal documents (e.g. example submissions)	• Press and media • The open call • Website • Internal documents (e.g. meeting reports, example submissions)

4.3.3 Case Study Results

The case study results are structured according the two investigated metrics, adequateness and perceived usefulness, which are subsequently presented in this section.

Adequateness of the Model

To report on model adequateness, we graphically represent the project activities of the two cases using the model as a baseline. This highlights not only the similarities but also the differences between our model and the investigated projects. Figures 4.3 and 4.4 summarise the activities of the CT project and the LDC project respectively. To increase readability, the figures represent the similarities in normal font; differences in *italic font*; and sub-activities in smaller font size.

Based on these graphical representations, we observe high adequateness of the model components. Both representations show strong concordance between the model components and the projects' activities. Examples include the strong

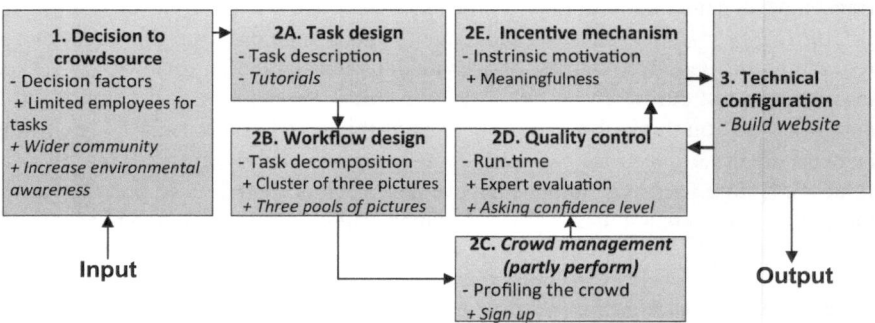

Fig. 4.3 Activities of crowd tagging (CT)

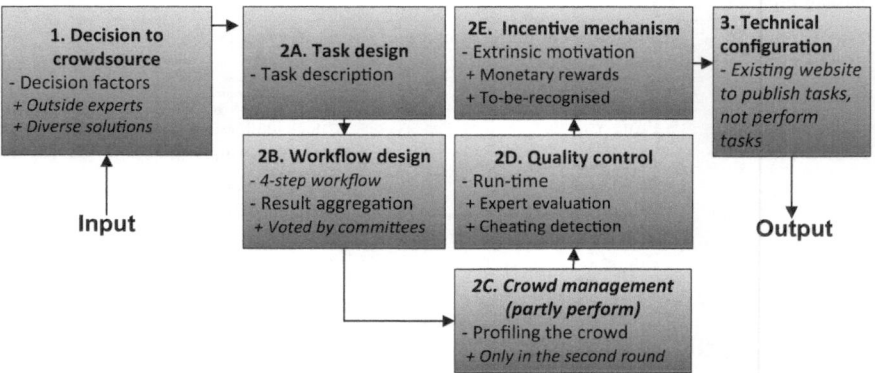

Fig. 4.4 Activities of logo design contest (LDC)

alignment on the decision to crowdsource, task design, workflow design, incentive mechanism, quality control, and partial alignment on crowd management and technical configuration. Several project sub-activities are also aligned with the model. However, both cases reveal several additional (sub) activities that are necessary to instantiate the components in practice. Examples include developing a tutorial in the task design of the CT case, and aggregating results through voting in the workflow design of the LDC case. Nevertheless, we find a strong alignment between the model components and the two projects, which suggests high adequateness of the model.

Specifically regarding the interdependencies suggested by the model, the two investigated projects are also largely aligned, i.e. they generally adopt the sequence of steps from input, decision to crowdsource, several aspects of crowdsourcing design, configuration, and finally to output. This alignment is stronger in the LDC case where most components follow the model sequence. In the CT case, we find strong alignment in the first four components, but some differences in the relationships among the last three components. More precisely, the three last components of CT were developed in a more iterative way, rather than following a sequential relationship. More details about the activities and their interdependencies are presented below.

Crowd Tagging (CT)

The CT project started with an input consisting of a large number of pictures to be analysed. To process these pictures, the project manager decided to adopt crowdsourcing. He stated three supporting reasons: (1) limited human resources to process the vast amount of data; (2) allowing the wider community to access the collected data; and (3) increasing environmental awareness of the community. While the later reasons are specific to the nature of CT as a citizen science project, the first reason, considered as the most important factor by the project manager, is consistent with the 'decision to crowdsource' component of the process model. More precisely, we consider the lack of internal employees to perform tasks as a factor driving the decision to crowdsource (Afuah & Tucci, 2012; Malone et al., 2010). Another reason CT should and did use crowdsourcing is the nature of the tasks. More precisely, tasks in CT were Internet-based; did not require interactive; were not confidential, and were partitionable. Thus, they are appropriate to crowdsource (consistent to Fig. 4.2).

After deciding to crowdsource, the project manager specified the crowdsourcing process itself, starting with task design. A task description was developed to promote the general aims of the project and explain how the task could be fulfilled by the crowd: "this research aims to evaluate the use of remote cameras to estimate abundances of non-native predators in urban environments. You will be shown a series of images, taken earlier this year, from various cameras placed around the Wellington city and asked to identify the animal in the photograph" [CT, Website]. The task design is consistent with the model component 2A. We also note the project included a tutorial and a visual explanation of the task, which served to train the crowd on how to perform the tagging. Such focus on training seems appropriate for this type of task, and the literature suggests that training the crowd may improve the results (Park, Shoemark, & Morency, 2014).

The CT project designed the crowdsourcing workflow through task decomposition. First, the whole activity was divided into sub-tasks of tagging three pictures, which the project alluded to as a cluster. This clustering was directly related to how data were collected in the project: "the camera takes three pictures every time they detect something. Thus, the group of three pictures helps make the task easier to perform" [CT, Project manager]. The project also divided the whole set of pictures into three pools: sign-up pool, working pool, and finished pool. The first pool included 20 clusters (of three related pictures), and the person who just signed up would start tagging the clusters in this pool. After a user finished ten clusters from the sign-up pool, the website would direct the user to the working pool. This pool included the remaining pictures that needed to be tagged, and thus was the main working zone. When a cluster had been tagged more than three times, it was considered finished and was moved to the finished pool. This pool stored the tagging results. While the three-pool decomposition is expected to improve reliability as seen below, we note that this decomposition can, and should, be extended for training purposes. More precisely, the first group can be used as gold standard data to give instant feedback and explanations as to why the crowd submissions may be (in)correct. By doing so, the crowd can learn and possibly provide better performance (Le, Edmonds, Hester, & Biewald, 2010).

According to the proposed model, crowd management aims at understanding the targeted crowd, which enables the assignment of tasks to suitable individuals to improve performance (Allahbakhsh et al., 2012; Khazankin et al., 2012b). The CT project manages the crowd by collecting users' information and evaluating their confidence levels on task performance. Collecting demographic information about the users was done at sign-up, which was required before a user could perform a task. More importantly, the project also managed the confidence levels by using two methods. The first method was based on the first pool with known answers for the tagging pictures. By comparing users' tags with the known answers, "we can say how reliable the users are". [CT, Project manager]. Another method asked the users directly how confident they are about their submissions in order to manage the confidence levels.

Since tagging was performed by voluntary users, there was no guarantee that the results would be of high quality. Thus, quality control seems necessary for projects similar to CT (Allahbakhsh et al., 2013). However, the CT project seems to have been limited in its quality control, comparing to what were suggested in the BPC model. CT was mainly based on expert evaluation after receiving tags from the crowd. This approach led to two concerns. First, this evaluation will heavily depend on the opinion of evaluator, as seen via "I see what the people say and what I say" [CT, Project manager]. Another issue was the large amount of data that needed to be evaluated; and the project currently does not yet address this issue but sees it as future work.

To attract the crowd, the project manager considered both extrinsic and intrinsic incentive mechanisms. Regarding the former, the project manager initially thought about providing vouchers to a popular, local wildlife sanctuary (Zealandia). However, he finally decided not to do so as he believed the users would be keen enough to contribute to the citizen science project anyway. As a result, the project was mainly based on intrinsic incentives. Similar to other citizen science projects (Brabham, 2012), this project suggests meaningfulness as an altruistic contribution to science, as stated in the website "every image you tag will help us to better understand the relationships between New Zealand's invasive mammals and native species".

In its technical configuration, CT built a crowdsourcing website that allows broadcasting the open call. This website also functioned as a platform, which enabled users to tag the pictures. CT decided to build its own website, rather than using some existing platforms, since the project members wanted to have full control over the whole set of crowdsourcing activities.

Logo Design Contest (LDC)

In LDC, the decision to crowdsource was based on two main factors: diverse solutions and external participants. The main reason for choosing crowdsourcing was the ability of the crowd to provide diverse and innovative solutions, as summarised by the project coordinator: "the university has decided to conduct the open contest to find ideas that are 'standard' [i.e. meeting the requirements] and creative". This is consistent with other crowdsourcing cases where external contributors can bring unique and innovative ideas (Brabham, 2010; Leimeister et al., 2009). Another factor influencing the decision to crowdsource was to utilise design contributors from outside the university. As logo design can be seen as a complex task (Schenk & Guittard, 2011), a certain level of expertise is necessary to generate a good design. Interestingly, saving costs (compared to hiring experts) was not considered as an important factor in the decision to crowdsource.

A key activity in crowdsourcing is task design (Model component 2A). Task design in LDC was presented through the announcement that was published on the University website and the local press. This announcement included the requirements for the logo, terms and conditions to join the contest, the submission deadline, and the prizes. Within these elements, the requirements played an important role as they specified what the solution should look like (Zheng et al., 2011). This considered two aspects: meaning of the logo and technical requirements. Meaning requirements were that the designed logo should represent the spirit of the University. The technical requirements specified, for instance, how many pixels were needed and the length of the slogan. We noted that while the technical requirements were specific, the meaning requirements were quite abstract. On the one hand, this abstraction left plenty of room for creativity in the design solutions. However, on the other hand, it did not fully show what the University board desired about the solution, which led to an extension of the contest because of several queries for clarifying the requirements [LDC, Project Coordinator].

The workflow design was an interesting activity with two distinctive aspects. First, while the model, consistent to Afuah and Tucci (2012), suggested task decomposition, LDC did not crowdsource decomposed tasks, but the whole logo design. This can be explained by the nature of logo design, which could be difficult to break down into smaller tasks. Additionally, crowdsourcing a whole task has been successfully adopted for several design contests, including bus stop shelter design (Brabham, 2012) and T-shirt design (Howe, 2006b). Second, LDC published its workflow in the open call. According the LDC announcement, the project workflow consisted of four steps: the crowd designs and submit their solutions; a preliminarily evaluation is conducted by the board; a short-list of submissions is chosen and given feedback, based on the board evaluation; and the final submissions are evaluated, ranked, and awarded. This provides transparency to the participants when explaining to them what will happen during the project.

The crowd management, which is specified in the model as task assignment and profiling the crowd, was not a focus in LDC. The project did not match the task to any specific members. Another aspect of the crowd management, which includes profiling the crowd (Allahbakhsh et al., 2012), was only processed in LDC when submissions were chosen for the second round. This was considered a limitation of LDC: "the management of crowd information was limited, which might be because we did not specify rules about providing information" [LDC, Project Coordinator]. As part of the crowd management, LDC had some communication with the contestants who wanted to find out more about the requirements. From the contest point of view, this kind of communication should be limited as it may create advantages for those contestants. Instead, a 'Q&A' section on the website, similar to the one deployed by Threadless (2015), should have been used.

To control quality, the LDC project used expert evaluation (Zhao & Zhu, 2014). In particular, the committee for aggregating results were also the evaluators, who assessed the submission quality and provided feedbacks. Since the number of submissions was not large (68 submissions), the use of a committee was a feasible approach. The project found a few cheating submissions that were likely copied from other logos. These submissions were mainly identified by the external experts who were experienced with logos and logo design contests [LDC, Project Coordinator].

To attract participants, the project used mainly extrinsic mechanisms, which consisted of monetary rewards and recognition by others. Like other contests, the monetary rewards were only provided for the winning solutions, which, in the LDC case, were two creative prizes and one final winning prize. The creative and winning prizes are quite valuable, equivalent to one and five month's salary of a typical office worker, respectively. Another motivation for the participants was that the project announced the winners on the University website, which is aligned with the to-be-recognised motivator (Brabham, 2012). Both of these motivations were clearly presented in the open call.

The technical configuration was rather simpler in this project, as LDC only used the website as a channel to publish the task and used emails to receive the submissions. This was because the project members were not aware of existing platforms/websites that can support crowdsourcing contests [LDC, Project Manager].

Overall, the results from the two cases confirm the adequateness of the proposed model to structure the project activities. Indeed, the two cases reveal a high alignment between the project activities and the model components. Adequateness is further confirmed in the interviews. The interviewees, when we show them the graphical representations of the project using the model, suggest these representations capture their projects activities. This quote evidences the suggestion: "we may miss some of the points, but we touch all of them" [CT, Project manager]. With the high adequateness, we expect that these members have a positive perceived utility of the model, as confirmed next.

Perceived Utility of the Model
Examining the perceived utility of the model, we interviewed the project members about the model, its components and sequence. The results were that all interviewees found the model to be a useful tool for structuring the crowdsourcing projects. This are demonstrated by the following comments.

> I think it will be nice to follow the model. [...]. Yes, I want to use the model, following this flow or at least have something to follow [CT, Project manager]

> The model is very well constructed and all of its activities should be necessarily for the project [LDC, Coordinator]

> As I said, I think this model is totally suitable. There is only slightly different on its progress, yet the meaning and purpose are similar. The approach and the steps are also similar [LDC, Project Manager]

Finding the usefulness of the model, these participants were extremely enthusiastic about applying the model for the future crowdsourcing projects:

> I think that any future crowdsourcing projects should apply strictly these steps, which will create better results [LDC, Coordinator]

> From my opinion, the model can be suitable for many activities that need the resources from the crowd [LDC, Project Manager]

In the model construction, we classified its components into plan-time and operation-time. It is interesting to find that the same idea was corroborated by a project manager. When we showed him the graphical representation of the model, he grouped the activities of the LDC project into planning and implementation, and states that:

> The component 2A and 2B [in the model] are similar to the planning phase of the project. The other components, including 2C, 2D, 2E, and 3, are implementation [LDC, Project Manager]

These comments expressed an agreement over the perceived usefulness of the model. Furthermore, the interviewees were curious to apply the model to future projects. Interestingly, when we discussed what aspects of the model are most useful, we found slightly different views between the project manager and coordinator roles. For instance, in the LDC case, while the project manager viewed the model as a tool for making decisions and management, the project coordinator instead stressed the role of the model in supporting communication among project members and in achieving a consensus. These differences suggest that usefulness can be perceived from different angles. Through this point, we highlight that if different roles can generate different insights when using the model, then the model's utility is expanded.

In summary, we conducted two case studies evaluating the process model. The results of the case studies found strong evidence that the model can represent the key activities of crowdsourcing projects. Furthermore, we also obtained evidence of the perceived usefulness of the model, inspired by the reception of the crowdsourcing experts. Consequently, we suggest that the proposed model addresses most organisational concerns within the crowdsourcing process, and that the model can be useful to support crowdsourcing projects.

4.4 Summary and Discussion

To guide organisations in their establishment of BPC, this chapter developed a conceptual model allowing organisations to understand the main building blocks of BPC. Using the identified building blocks extracted in the previous chapter, we constructed a process model of BPC consisting of seven components. The construction was based on the 'wisdom of researchers', which enabled us to build the model faithfully representing BPC. The model was evaluated using the case study approach. Two real crowdsourcing projects were used for this evaluation. The results indicate that the model is adequate and useful in structuring the main crowdsourcing activities.

Overall, the model represents the main structures of BPC to support the establishment of crowdsourcing as an organisational business process. It provides a broad view of what activities that organisations need to be considered when planning, designing and instantiating crowdsourcing processes. This broad view, on the one hand, overcomes the excessive ad hoc criticism complained in the crowdsourcing literature (Geiger & Schader, 2014; Mao et al., 2017). On the other hand, it represents only the abstract view but not the deconstructed view, both of which together characterise BPC. From the deconstructed view, the process model and its components need to be further analysed into detailed elements. The following chapter addresses this need, which builds an ontology from both abstract and deconstructed views.

Chapter 5
Ontology of Business Process Crowdsourcing

An ontology explicitly defines the terms in the domain and relations among them, which with to represent knowledge.
—Adapted from Gruber (1993)

The purpose of this chapter was to examine BPC from both abstract and deconstructed views. For this purpose, we built an ontology of BPC that captured main concepts and relationships in the domain (Corcho et al., 2003). As a result, the BPC ontology provided the ontological structure necessary to understand the main constituents of BPC. Furthermore, ontologies can enhance reasoning knowledge (Valaski, Malucelli, & Reinehr, 2012), and thus constructing a BPC ontology made a further step towards supporting BPC establishment. Keeping the foundations of design science research, we built and then evaluated the ontology. Specifically, we built the ontology following the conceptual model and using raw knowledge materials from the previous chapters. We evaluated the ontology by comparing it with an ontological version generated by software. We note that some parts of the ontology were presented in our conference paper (Thuan et al., 2015).

5.1 Ontology Building Process

To start the building activity, we reviewed the ontology engineering literature to identify and justify the activities of ontology construction. This led us to adopt the two activities commonly used in ontology engineering: *ontology capture* (Uschold & King, 1995) and *knowledge organisation* (Küçük & Arslan, 2014).

The ontology capture
This activity aimed at deriving ontological elements. We analysed the knowledge sources extracted by our scoping review for concepts, hierarchical relationships, decision-making relationships, and business rules related to crowdsourcing. This analysis was supported by the following ontology schema, which provided a structured meta-model for knowledge source analysis (Levy & Ellis, 2006; Okoli &

© Springer International Publishing AG, part of Springer Nature 2019
N. H. Thuan, *Business Process Crowdsourcing*, Progress in IS,
https://doi.org/10.1007/978-3-319-91391-9_5

Schabram, 2010). The ontology schema is graphically presented in Fig. 5.1. It consisted of two main areas named *ontological representation* and *knowledge representation*, which were used to analysed the knowledge sources in both deductive and inductive approaches.

In the left-hand side of Fig. 5.1, the ontological representation included four ontological elements: concepts, hierarchy relationships, decision-making relationships, and business rules. While concepts and hierarchy relationships were important to structure a domain knowledge (Corcho et al., 2003; López, Gómez-Pérez, & Corcho, 2004), decision-making relationships and business rules provided reasoning knowledge and thus were also critical when establishing business process crowdsourcing. The ontological representation was mainly used for deduction.

When deducting, we analysed the sources for concepts and sub-concepts using the components of conceptual model as the pre-defined themes. For each extracted concept/sub-concept, we specified its name, synonym, and description. Since ontologies include both the concepts and the linked extensions between concepts (Corcho et al., 2003; López et al., 2004), the next considered element was the relationship. In this element, we analysed not only hierarchical relationships but also decision-making relationships. The former referred to taxonomic structures in the domain (López et al., 2004), where we adopted five hierarchy relationships commonly used in ontology engineering, including 'is a', 'include', 'categorise', 'instance of', and 'based on'. The latter referred to reasoning relationships supporting decision-making. In this type of relationship, we chose the following ones: 'positively influence', 'negatively influence', and 'associate', which were popularly suggested in the literature (e.g. Chandler & Kapelner, 2013; Hoßfeld et al., 2013). The last considered element concerned business rules, which added constraints to the concepts and relationships.

In the analysis, we faced an issue that some emerging elements did not align with our pre-defined codes. This issue is normally expected in emerging research fields like crowdsourcing where diverse views and methods are adopted (as discussed in Chap. 2). Addressing this issue, we used an inductive analysis that

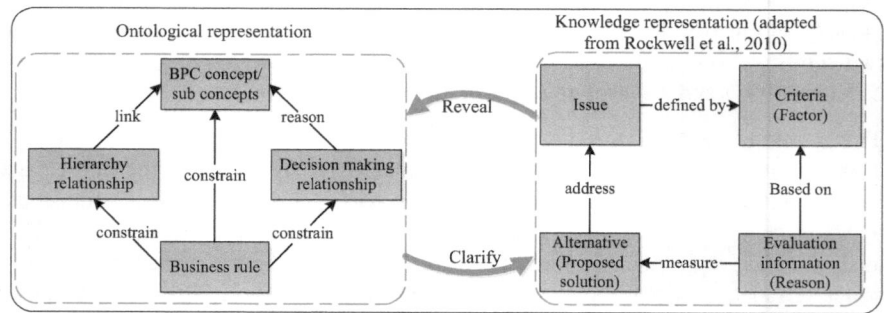

Fig. 5.1 Ontology capture schema (adapted from Rockwell et al., 2010)

allowed studying and characterising knowledge from the bottom up. Thus, we adapted the knowledge representation approach proposed by Rockwell et al. (2010) (the right-hand side of Fig. 5.1), which characterised knowledge by four categories: issue, criteria, evaluation information, and solution alternatives. Based on these categories, we applied the following questions when analysing the knowledge sources: what are the main issues related to BPC? How can these issues be defined, i.e. characterised by what factors? What alternatives can be chosen to address the issues? And how can we evaluate the proposed alternatives?

We note that the two aspects of the ontology schema (ontological representation and knowledge representation) support each other, and together assist the data analysis. While the ontological representation allows eliciting existing knowledge, the knowledge representation helps further clarify and fulfil the knowledge gaps. Another strong point of this schema is its ability to be used in both deductive and inductive approaches. Whilst the deductive analysis ensures that the captured elements align with the conceptual model and thus address core concepts of the crowdsourcing domain, the inductive analysis captures the emerging nature of the crowdsourcing field. This schema was applied to every source in our pool. As a result, we captured a large number of ontological elements: concepts, hierarchy relationships, decision-making relationships, and business rules, which are structured in the next step.

The knowledge organisation
This activity synthesised the ontological elements extracted in the previous activity and organised them to form the ultimate BPC ontology. The 'wisdom of researchers' was again applied to each ontological element: distilling concepts, hierarchy relationships, decision-making relationships, and business rules. This allowed finding salient elements that were supported by multiple sources of knowledge. We noted that during synthesis, there were several cases where the sources were inconsistent on certain extracted elements. For instance, different reviewed sources proposed different hierarchical relationships for quality control mechanisms. Some structured quality control into design-time and run-time mechanisms (Allahbakhsh et al., 2013). A slightly different categorisation defined before-task, during-task, and after-task mechanisms (Alonso, 2013). Other authors proposed completely different categories, including supervised and unsupervised mechanisms (Baba & Kashima, 2013). In these cases, the 'wisdom of researchers' helped to choose the elements supported by the majority of sources.

Based on the synthesised elements, we then organised and structured them into the ontology. Since the relationships revealed the fundamental structure of the BPC domain, the ontology ended up being organised around them. It was nevertheless important to note that the organisation process was highly iterative, where we followed a trial-and-error process and updated the ontology several times. The organisation process was also performed using inductive and deductive strategies. This could be exemplified with the procedure of obtaining hierarchical relationships. In the deductive synthesis, we followed guidance from the relationships suggested by a majority of the reviewed sources, e.g. quality control can be classified into design-time and run-time mechanisms (Allahbakhsh et al., 2013; Alonso, 2013). However, no guidance was found for some groups of concepts. In these cases, we aligned with Nickerson et al. (2012) and inductively identified common characteristics of the (sub) concepts and then proposed their classification schema.

5.2   An Ontology of BPC

We now report the results from the ontology building. The section starts with the most popular concepts of BPC, and then summarises the hierarchy relationships that are used to organise the ontology. Subsequently, decision making relationships and business rules in the domain are described. Here, we provide further details regarding salient concepts, hierarchy relationships, and decision making relationships.

5.2.1   Salient Concepts

Adopting the 'wisdom of researchers', we focus on concepts suggested by multiple knowledge sources. Table 5.1 presents the 39 most salient (sub) concepts supported by at least 10 sources. At a high level, Table 5.1 represents the main building blocks of BPC, while at a more detailed level it clarifies these building blocks with their categories and sub-concepts. This clarification suggests that the conceptualisation captured in Table 5.1 has a more detailed level of abstraction compared to Table 3.2. To increase readability, Table 5.1 shows several building blocks of BPC in **bolds**, categories in *italic*, parent concepts in Capital-first-letters, and sub concepts and attributes in all-lower-letters.

In Table 5.1, the concepts were supported by at least ten knowledge sources, which indicated that they were consensually salient in the BPC domain. Thus, they were treated as the core elements of the ontology. Besides these core elements, we noted that other concepts were still considered in the ontology construction. The reason was that the importance of a concept should be indicated by not only the number of supporting sources but the relationships with other concepts, given the important roles of relationships in ontologies (Guarino, Oberle, & Staab, 2009; Sánchez & Moreno, 2008).

5.2.2   Hierarchy Relationships

Using the concepts identified in the previous section, we organised them hierarchically to give a more holistic picture on the BPC domain. Our analysis disclosed a

Table 5.1 Salient concepts of BPC

Concept	Number of supporting sources
Quality control	**69**
Design-time	11
worker selection	16
Run-time	13
identifying malicious behaviour	19
gold standard	16
output agreement	12
Incentive mechanism	**46**
monetary reward	29
fun	11
Crowdsourcing output	**38**
output quality	36
Task design	**37**
Task description	10
Crowd management	**34**
Task assignment	20
Profiling the crowd	10
worker profile	10
worker reputation	10
Crowdsourcing task	**34**
simple task	13
complex task	12
Decision to crowdsource	**26**
Decision factor	19
Task characteristic	30
ease of task delineation	13
partitioned task	11
Availability of the crowd	19
Risk & Challenge	16
Availability of crowdsourcing platform	10
Characteristics of the crowd	**23**
Type of worker	12
Motivation of the crowd	10
Workflow design	**21**
Result aggregation	29
Task decomposition	10
Control and feedback	**17**
Technical configuration	**14**
Platform (intermediary)	13

Name conventions
Building blocks are in **Bold**; categories are in *italic;* parent concepts are in
Capital-first-letter; and sub concepts or attributes are in all-lower-letter

diverse hierarchical structure of BPC, which can be seen by counting each type of relationship: 'include' (78 sources), 'instance of' (30 sources), 'categorise' (22 sources), 'based on' (22 sources), and 'is a' (19 sources). To organise the identified relationships and concepts in a manageable way, we followed a trial-and-error process, in which we tried different structures, including a tree structure, a network structure, and a layer structure. Yet, the first two structures did not appear to be suitable to our goals. The tree structure consisted of several branches with division concerns, and thus was different from the holistic view of BPC. While supporting a holistic view, the network structure made the ontology representation too complex, with many links and crosscuts.

As a result, we adopt the layered structure, which diminishes complexity by arranging concepts and relationships into layers. Furthermore, the layered view is appropriate for integrating crowdsourcing into the organisation as integration is usually done at different levels (layers) of concerns (Giachetti, 2004; Hasselbring, 2000). The layered structure is presented in Fig. 5.2. In this figure, the three kinds of relationships that are represented are 'include', 'categorise', and 'based on'. We note that the 'is a' relationship has been transferred to 'include' and 'categorise' relationships, and the 'instance of' relationships are not shown in order to reduce the complexity and to increase readability of the figure.

Figure 5.2 represents BPC main concepts and hierarchical relationships of BPC, which according to Corcho et al. (2003) captures a lightweight ontology of BPC. The ontology is organised in four layers with an increasing level of detail from inner to outer. At the heart of the framework, *the core layer* represents main concerns that should be focused in order to establish BPC. This layer captures the most abstract building blocks of BPC, which are aligned with the conceptual model (Chap. 4). To clarify these building blocks, the other concepts identified in Table 5.1 are further organised. They are presented in the next layers as: process layer, data layer, and data attribute layer, which is one typical schema for classifying IS objects (Giachetti, 2004; Zachman, 1987).

As seen in Fig. 5.2, *the process layer* describes plans of action that are performed in particular building blocks. In other words, the layer details each building block through a set of activities addressing a particular concern in BPC establishment. As a result, the process layer links with the core layer mainly through the 'include' relationship. For instance, the workflow design includes three activities: identifying type of task (Dai, Lin, & Weld, 2013), task decomposition, and results aggregation (Kittur et al., 2013). From an IS perspective, most processes/activities require certain information and data to be used (Berente, Vandenbosch, & Aubert, 2009), which suggests a strong link between this layer and the next data and data attribute layers.

The next layer is the *data layer* that shows data entities and information used by the activities. For instance, to examine factors of the decision to crowdsource, decision makers need to process four data entities about task characteristics, people, management, and infrastructure, as already discussed in Sect. 4.2. Regarding the relationships between the process and data layers, the data usage is denoted through two main types of relationships: 'include' and 'based on'. Some activities clarify (or

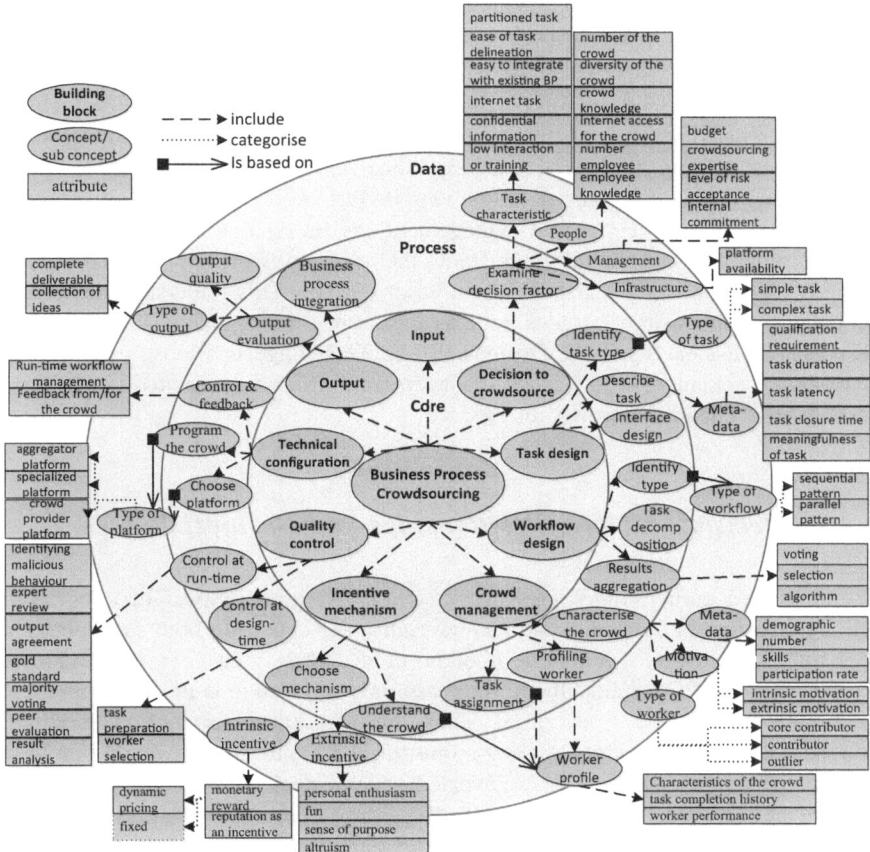

Fig. 5.2 A lightweight ontology of BPC

include) data entities, e.g. 'characterising the crowd' includes clarification of 'meta-data'. Other activities are founded on pre-defined data, e.g. the activity 'identify type of workflow' is based on data about 'type of workflow'.

The outermost layer then represents the *attributes* of each data entity. This representation is expressed through 'include' and 'categorise' relationships. The 'include' relationship shows that an attribute is a part of the data entity. For example, the meta-data for task description includes qualification requirement, task duration (Chilton, Horton, Miller, & Azenkot, 2010), and may also consists of other attributes. Also presenting the whole-part relationship, the 'categorise' relationship further requires that all of the attributes make up an exhaustive decomposition of the whole concept. For instance, the reviewed sources suggest three categories of workers: core contributor, contributor, and outlier (Chanal & Caron-Fasan, 2010; Stewart et al., 2010).

From the hierarchical relationships to structure the lightweight ontology, we note here two interesting points. First, these relationships enable explicitly structuring the related (sub) concepts in the domain. For instance, Fig. 5.2 shows that quality control can be managed at both design-time and run-time, each of which includes several detailed mechanisms for quality control management. Second, by ontologically structuring these relationships, some interesting links that were not revealed by individual sources are shown in Fig. 5.2. The connection between incentive mechanism and crowd management can be seen as an example. In particular, organisations should understand the targeted workers when designing incentive mechanism (Chanal & Caron-Fasan, 2010). This understanding can be achieved through worker profiles built by the 'crowd management' building block (Khazankin, Psaier, Schall, & Dustdar, 2011). This suggests a close link between 'incentive mechanism' and 'crowd management', which is presented via the concept of 'understand the crowd' in Fig. 5.2.

5.2.3   Decision Making Relationships and Business Rules

As mentioned earlier, ontologies can be classified into lightweight ones representing structured knowledge and heavyweight ones capturing both structured and reasoning knowledge of a domain (Corcho et al., 2003; Valaski et al., 2012). For the purpose of BPC establishment, the reasoning knowledge is important as it can guide the establishment activities. Thus, this section aims to add reasoning knowledge to the lightweight ontology, thus turning it into a heavyweight ontology.

For this purpose, our analysis revealed several decision-making relationships and business rules in the BPC domain. Regarding decision-making relationships, the reviewed sources identified a number of relationships, including 89 'positive influences', 17 'negative influences', and 10 'associations'. Again, the "wisdom of researchers" was applied to choose the relationships either suggested by multiple sources or linked salient concepts. We then organised the chosen relationships based on the lightweight ontology but removed the process layer since only a few decision-making relationships could be identified in the layer. For simplification, Table 5.2 summarises the association relationships while Fig. 5.3 represents the other relationships.

Figure 5.3 presents the key decision-making relationships in the BPC domain. Besides the 'include' relationships adopted from the lightweight ontology, there are two main relationships: positive (P) and negative (N) influences. The numbers, shown either on the line or behind the concept, indicate how many sources supporting a particular relationship. For instance, there are three reviewed sources suggesting that quality control positively influences output quality. In addition to the main relationships, Fig. 5.3 also notes some cases where the reviewed sources do not find statistical evidence to support a particular relationship, which is presented as 'does not influence'.

Table 5.2 Association relationships in BPC domain

Concept 1	Relationship	Concept 2
Output quality	Associate	Worker profile, task design, quality control, task complexity, and monetary reward
Type of task	Associate	Type of workers, incentive mechanism, benefits for organisations, task design, and result aggregation
Incentive mechanism	Associate	Type of worker
Task design	Associate	Quality control

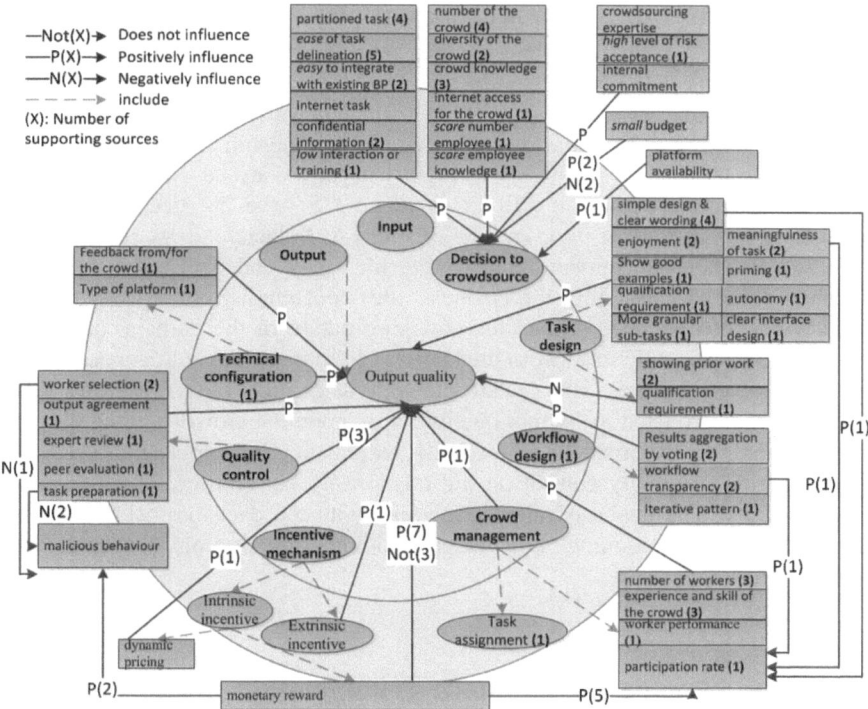

Fig. 5.3 Positive and negative influence relationships

As represented in Fig. 5.3, our analysis reveals three important findings. First, although there are diverse relationships between different concepts, the key one is how to influence crowdsourcing 'output quality' (Archak, 2010; Chandler & Kapelner, 2013). This is logical as 'output quality' indicates the success of crowdsourcing projects. Furthermore, a crowdsourcing project is effective only if it

Table 5.3 Some examples of business rules

Example of business rules
For taxonomy creation, *output quality* of crowdsourcing is equivalent to 80–90% of expert output (Chilton, Little, Edge, Weld, & Landay, 2013)
The more redundancy in performing a task (in iterative workflow), the better *outcome* of task results (Karger, Oh, & Shah, 2013)
Without quality control, more crowd workers are needed to achieve the same level of *output quality* (Tetreault, Chodorow, & Madnani, 2014)

can achieve high quality output. As a result, we have allocated the 'output quality' concept in the centre of Fig. 5.3. Second, a few conflicting relationships can be found, such as the influence of monetary reward on output quality. While seven sources suggest a positive influence, three other sources do not find statistically significant results to support the influence. In these cases, further research is necessary to test and confirm the relationship. Finally, we found that the number of sources supporting a particular decision-making relationship is rather low (mainly from one to three). This lowness is logical as the analysed sources consist of academic articles/papers in the IS field, where "IS have not been interested in publishing replications of prior studies" (Dennis & Valacich, 2014, p. 1).

Based on the foundation offered by the hierarchy and decision-making relationships (Figs. 5.2 and 5.3), the business rules constraining concepts in the BPC domain has also revealed. Given the emerging nature of the domain, only a few business rules were extracted from the knowledge sources. Table 5.3 presents some examples of the business rules related to 'output quality' only. The rules define three constraints related to 'output quality': how good the crowdsourcing output is in comparison to experts'; the role of task redundancy in output quality; and the moderate role of quality control on the relationship between crowd workers and output quality. The business rules, combining with the decision-making relationships, form the reasoning knowledge guiding organisations in their BPC establishment.

5.3   BPC Ontology Evaluation: Triangulation

This section discusses the evaluation of the proposed ontology. According to Venable et al. (2012), evaluation helps: (1) assess artefact's utility as to whether it achieves the stated purpose; (2) compare the built artefact against other artefacts; and (3) consider side-effects and weaknesses of the artefact for future improvements. This book chose to evaluate the ontology using the second approach for two reasons. First, this type of approach, compared to the other approaches, fits with the constrained resources of the current research project. The other approaches evaluate whether the artefact works and how to improve it, which needs considerable efforts on practical, long-run applications. According to Gregor and Hevner (2013), these approaches may not be feasible in research projects with limited resources like the current case. Second, the use of different evaluation methods in a design science project has been widely suggested (Sonnenberg & vom Brocke, 2012a; Venable, Pries-Heje, & Baskerville, 2016). As the previous research chapter has already used the case study evaluation and the following chapter will use experiments and focus groups, the use of triangulation in the current chapter can provide a complementary evaluation. As a result, triangulation, where we compared different versions of the ontology, was chosen for this evaluation.

Having adopted triangulation to evaluate the BPC ontology, we considered two metrics popular to assess ontology: clarity (Akdemir, Turaga, & Chellappa, 2008; Fan, Hua, Storey, & Zhao, 2016) and coverage (Fan et al., 2016; Shanks, Tansley, & Weber, 2003). We defined *clarity* as 'the degree to which the ontology clarifies concept meanings and reduces ambiguity in the domain', and *coverage* as 'the level that the ontology covers the semantics in the domain'. Using these two metrics, we compared our ontology with an ontological version generated by software.

We developed an automated version of the ontology using the same sources of information. This version was built automatically using a software tool—OntoGen that generates ontologies from text (Fortuna, Grobelnik, & Mladenic, 2007). While our ontology building was based on a detailed review of BPC sources, the auto-mated ontology was built using the abstracts of the same sources. This is because abstracts are expected to consist of key concepts, relationships, and findings of the sources. Furthermore, the use of abstracts for ontology building is suggested by Vogrinčič and Bosnić (2011) regarding the use of article abstracts for their ontology construction. When using OntoGen to process the abstracts of BPC sources, the outcome is presented in Fig. 5.4.

We then analytically compared our own ontology with the automated ontology. We find a high consistency on the main ontological elements. When comparing Figs. 5.2 and 5.4, a strong match is noted for the core building blocks (i.e. tasks, quality control, incentive mechanism, technical configuration, and the crowd). Furthermore, several detailed concepts are also similar, e.g. intrinsic motivation and extrinsic motivation can be found in both figures. Despite a few differences, we find high consistent results between the two ontological versions. As a result, the comparison validates our ontology building process through triangulation as sug-gested by Carlsson et al. (2011) that "to strengthen the validity of design [theories], test triangulation may be beneficial" (p. 117).

We further consider the triangulation results according to the two investigated metrics, coverage and clarity. We find that the results confirm the high coverage of the ontology. More precisely, most of the concepts generated by OntoGen have already been captured by our ontology (comparing Fig. 5.2 with Fig. 5.4). Further, our ontology covers not only concepts, but also hierarchical and decision-making relationships, while the automatic version is quite limited regarding the type of relationships, e.g. excluding non-hierarchical relationships. These points indicate that our ontology has covered a wider range of semantics in the domain.

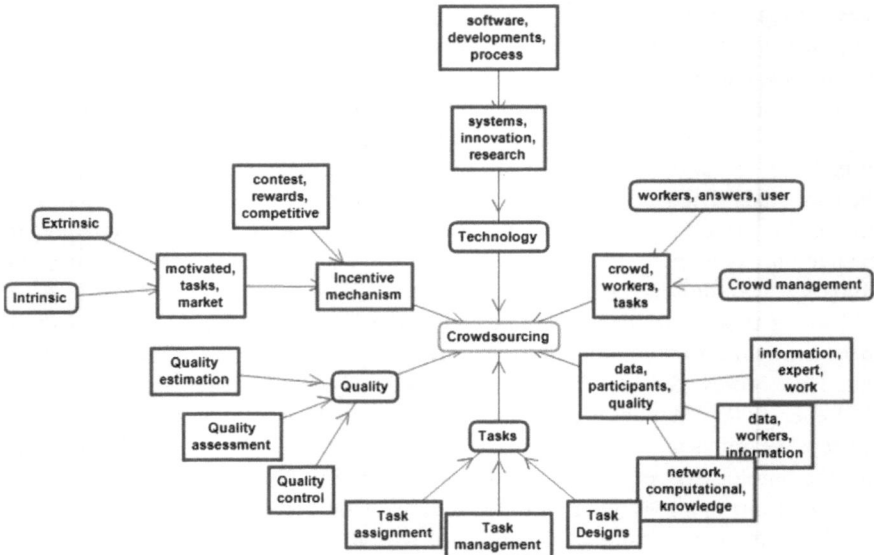

Fig. 5.4 Ontological version generated by OntoGen (OntoGen was developed by Fortuna et al., 2007)

Regarding clarity, our ontology advances the automatic version through two important points. First, our ontology provides clearer concept meaning. More precisely, since the automated approach groups and defines concepts based on the frequency of occurrence, rather than meaning, several extraneous composite concepts have occurred, e.g. the combination of network, computational, and knowledge. Such composite concepts do not provide clear meaning and thus make the automatic ontology harder to understand. In contrast, our ontology decomposes the composite building blocks into detailed concepts to make it easier to understand. Finally, our ontology distinguishes different types of concepts, e.g. building blocks, activities, data, and attributes, which is necessary for the related information system development. This capacity was not yet supported by the tool.

Given the discussion, the triangulation results allow us suggesting that the ontology highly covers and clarifies the domain semantics. Thus, we strongly believe that our approach, which builds the ontology through systematic analysis and organisation of BPC knowledge sources, is appropriate to construct the ontology of BPC. This appropriateness is supported by Osterwalder (2004) and Miah et al. (2009) regarding their similar approaches to construct domain ontologies in the IS discipline.

5.4 Summary and Discussion

This chapter built a lightweight and then heavyweight BPC ontology. We analysed and structured the BPC knowledge sources to identify the main concepts, hierarchical relationships, decision-making relationships, and (a few) business rules defining BPC. These elements were then organised into a lightweight ontology considering BPC building blocks, business processes, data entities, data attributes, and their hierarchy relationships. Then, decision-making relationships were added, which turned the ontology into a heavyweight ontology. To evaluate the ontology, we compared it with an automated ontology generated by OntoGen, which strengthens its validity (Carlsson et al., 2011). The results show high coverage and clarity of our constructed ontology.

Complement to the current triangulation evaluation, we recently assessed the ontology using an application-based evaluation. The detailed evaluation is presented in our journal publication, and interested readers are directed to Thuan et al. (2018). That evaluation results show the applicability of the ontology, actually implemented in a working system. The combination of the triangulation and application-based evaluations show high coverage, clarity, and applicability of the BPC ontology.

The role of the BPC ontology can be seen either together with the conceptual model or as a standalone artefact. Comparing to the conceptual model, the ontology provides a more detailed decomposition of BPC, decomposing the building blocks into detailed concepts and their relationships. As a result, the conceptual model and ontology can be used as two interrelated levels of BPC decompositions, which allow organisations analyse, plan, and deploy a business process based on crowdsourcing. As a standalone artefact, the BPC ontology provides a knowledge base that consolidates the domain knowledge. It structures key concepts, hierarchical structures, and decision-making relationships of the domain, from which knowledge can be interred. With this knowledge base, instantiated artefacts can be further constructed and developed to support BPC, which is the focus of the next chapter.

Chapter 6
Business Process Crowdsourcing: Decision Support Tool

*Decision support systems have been developed to facilitate
better decision making for difficult and complex structured,
semi-structured, and unstructured decisions.*
—Hosack et al. (2012)

This chapter constructed a decision tool supporting the establishment of BPC. This construction fulfils a need in the BPC domain and plays an important role in the research process. In the BPC domain, there is a need for decision support systems to address the complexity of BPC establishment. The complexity of BPC establishment has been revealed in the previous research stages and also emphasised by other researchers (Pavel Kucherbaev et al., 2013; Tranquillini et al., 2015). More precisely, BPC establishment involves not only the several stages deploying in the crowdsourcing strategy, but also several information structures supporting the establishment (as seen via Fig. 5.2). Given this complexity, it is necessary to help decision makers—managers and process designers—making analytical decisions in BPC establishment. Consequently, the construction of a decision tool supporting BPC establishment fulfils the need in the domain.

In this book, the role of this chapter is essential for both alignment with the other chapters and for itself. In alignment with the other chapters, this chapter articulates the knowledge, which was extracted in Chap. 3, conceptualised in Chap. 4, and ontologically detailed in Chap. 5, in order to build a decision tool regarding the BPC process. Consequently, it contributes to move forward the theoretical efforts from the previous chapters. By itself, this chapter has its unique outcome and position in the research process. The outcome of this chapter is a decision tool supporting BPC, which is clarified as an instantiation artefact per se (Hevner et al., 2004). Furthermore, this research is also important due to its ultimate position in the research process. This position implies that the success of tool development has a great influence on the success of the whole project. If we can demonstrate the utility of the tool in practice, this increases confidence for not only the tool construction but also the knowledge base built into the tool.

Given the important roles, the decision tool should be constructed and evaluated in a rigorous way. In the construction, we ensured rigour through a solid knowledge

© Springer International Publishing AG, part of Springer Nature 2019
N. H. Thuan, *Business Process Crowdsourcing*, Progress in IS,
https://doi.org/10.1007/978-3-319-91391-9_6

base and an appropriate development method. Regarding the former, we built an architecture that guided the development process. This architecture embraced the BPC ontology as its knowledge base module, which solidly supported DSS development (Delir Haghighi, Burstein, Zaslavsky, & Arbon, 2013; Miah, Kerr, & von Hellens, 2014). We also structured a set of decision tables based on the decision framework (Fig. 4.2), which were operationalised and embraced in the tool. Regarding the development method, we followed Lim et al.'s (2008) suggestion to adopt a rapid prototyping method. This method allowed managing rigour through iterative development, assessment and revision of a few prototypes (Kordon, 2002; Lim et al., 2008).

In the tool evaluation, a combination of two different evaluation techniques enabled rigour (Sonnenberg & vom Brocke, 2012a; Venable et al., 2016). We purposefully adopted two techniques with quite different natures: experiments and focus groups. The former is a quantitative, individual-based, and artificial evaluation, whereas the latter is a qualitative, group-based, and naturalistic evaluation. These two techniques allowed us to assess the tool not only by measuring the level of support that the tool can provide, but also by examining the participants' opinions on the usefulness of the tool. To further ensure rigour, this stage followed rigour guidance for conducting experiments suggested by Montgomery (2012), and focus groups suggested by Tremblay et al. (2010).

Based on the above discussion and in alignment between the decision tool with the conceptual model and ontology, we note that the tool should accomplish the following requirements:

- Help managers making the decision to crowdsource. We note that this decision is driven by multiple by multiple factors, as discussed in Sect. 4.2. The tool needs to operationalise these decision factors and should provide guidelines and recommendations to make the decision to crowdsource or not.
- Build a comprehensive integrated view of BPC, which is currently missing in the domain. In other words, the tool should support the integrated BPC process, not individual activities. The literature has suggested that such a comprehensive view can be reached by using sound domain ontologies (Miah et al., 2014; Osterwalder & Pigneur, 2004).
- Support micro-decisions related to the BPC building blocks and workflows (Fig. 5.2). That is, within each building blocks, the (sub) issues, their alternatives, and guidance to choose among these alternatives should be specified.
- As an instantiation artefact, the tool should provide a means for processing and presenting knowledge related to BPC establishment.

We note that this chapter is based on the journal publication by Thuan et al. (2018) with two extensions. First, we provide details of the tool construction, including the architecture, decision tables, pilot experiments, and the revision of the tool based on the pilot experiments. Second, we extend the tool evaluation with the focus group results. The qualitative evaluation by the focus group is important to complement the quantitative evaluation by the experiments.

6.1 Architecture and Decision Tables

DSS architecture defines key components of the DSS system. Though diverse types of DSSs have been developed, their abstract architecture seems fairly consistent. Holsapple (2008) summarises this consistence and suggests an overall architecture of DSSs consisting of four components: language component, problem-processing component, knowledge component, and presentation component. Given that Holsapple's (2008) suggested components are abstract and clearly separate the important concerns of DSSs, they were adopted to design the decision tool. We note that the term 'component' was already used to refer to the model components in Chap. 4. To avoid confusion, we refer below to the architectural components as modules.

The tool's architecture was based on Holsapple's (2008) abstract architecture, where the language and presentation modules were combined into the graphical user interface (GUI). Consequently, the architecture consisted of three modules: GUI, problem processing module, and knowledge base module, which are depicted in Fig. 6.1. *The GUI module* is responsible for the interaction between the tool and the users. It receives input parameters from the users, offers descriptions about the parameters, and provides related advice. *The problem process module* handles these inputs, where they are used to formulate the decision and the related context. This module also controls the flow of inputs by adapting what elements the GUI presents, which manipulates data entries based on the knowledge module. *The knowledge module* is based on the BPC ontology constructed in Chap. 5. Based on the ontology, the knowledge module can perform what-if analysis by comparing the domain knowledge with the input parameters. Consequently, the tool can identify inconsistencies in the inputs and provide advice on how to set up a BPC process for a particular organisational context, which in turn are presented as GUI's outputs.

In the construction of the knowledge module, we faced a challenge related to reasoning. In particular, there was a current lack of reasoning rules for making the decision to crowdsource or not (Zhao & Zhu, 2014). Addressing the challenge, we further examined the decision factors presented in Fig. 4.2 for actionable rules, which were structured by decision tables. Decision tables are suggested by Huysmans et al. (2011) as the most effective technique in terms of presentation and interpretability, compared to decision trees, propositional rules, and oblique rules. Furthermore, decision tables are easy to embed within computer software like the decision tool.

Decision Tables for Making the Decision to Crowdsource
This section presents a series of decision tables that provide actionable recommendations for making the decision to crowdsource. These tables, previously presented in Thuan et al. (2016), were drawn from the decision factors (Table 3.3) and decision framework (Fig. 4.2). Given the four layers of the decision framework, three decision tables were arranged according to the layers: task properties, people, and management. An exception is the environment layer that has only one

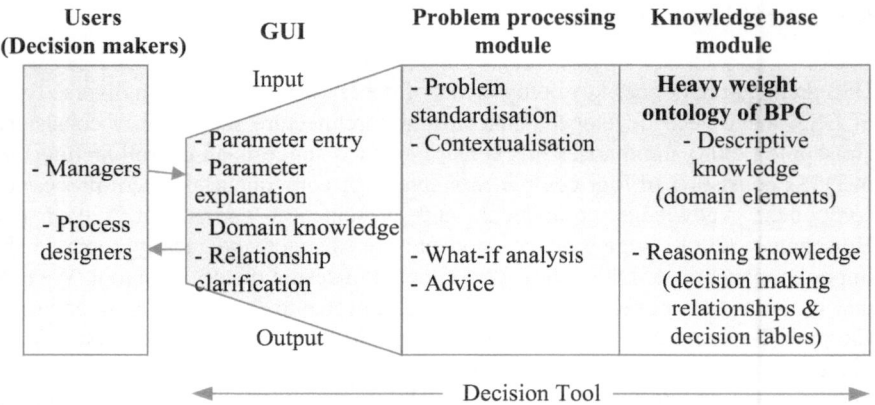

Fig. 6.1 Tool architecture (adapted from Holsapple, 2008)

decision factor, and thus does not need a separate table. Tables 6.1, 6.2 and 6.3 present the three decision tables.

Task properties with their central roles in the decision to crowdsource are presented in the first decision table (Table 6.1). From our knowledge base, the decision to crowdsource should only be made for tasks that satisfy three conditions: (1) can be performed through the Internet (Brabham, 2008a; Muntés-Mulero et al., 2013); (2) enable to integrate with the organisational business processes (Kittur et al., 2013; Sakamoto et al., 2011; Tranquillini et al., 2015); and (3) do not require many interactions (Afuah & Tucci, 2012; Burger-Helmchen & Pénin, 2010; Muntés-Mulero et al., 2013). In addition to these definite conditions, there are three other 'should-be' conditions. First, tasks should be well defined, which may be processed in the latter stages of the BPC process (Lloret et al., 2012; Muhdi et al., 2011; Zogaj et al., 2014). If tasks include confidential information, it is necessary to perform additional actions for hiding sensitive information (Feller et al., 2012; Lu et al., 2015). Finally, tasks that can be divided into small pieces of work have more chance to crowdsource (Afuah & Tucci, 2012; Malone et al., 2010). An exception is the case of crowdsourcing contests, where several tasks that are not necessarily divisible are successfully crowdsourced in the form of open contests.

Table 6.2 considers the roles of the availability of the crowd and organisational employees playing in the BPC process. Djelassi and Decoopman (2013) and Saxton et al. (2013) all agree on the availability of the crowd as a pre-condition for crowdsourcing. Without meeting this condition, the decision to crowdsource is inappropriate. The second condition considered in this table is the availability of internal employees. Following Afuah and Tucci (2012), we suggest that organisations should crowdsource in cases where they cannot allocated employees or neighboured agents to accomplish the tasks. In a similar vein, Lu et al. (2015) from a resource-based view advise "when a firm finds that its internal [human] resources and capabilities cannot satisfy the company's strategic objectives, the external

Table 6.1 Decision table: task properties

Conditions											
Internet: Yes (Y) vs. No (N)	N	Y	Y	Y	Y	Y	Y	Y	Y	Y	Y
Ease of integration with existing BP	–	N	Y	Y	Y	Y	Y	Y	Y	Y	Y
Interactive	–	–	Y	N	N	N	N	N	N	N	N
Ease of delineation	–	–	–	Y	Y	Y	Y	N	N	N	N
Confidential information	–	–	–	Y	Y	N	N	Y	Y	N	N
Partitionable	–	–	–	Y	N	Y	N	Y	N	Y	N
Actions											
Not to crowdsource	X	X	X								
Should crowdsource						X					
Crowdsource with additional action (CSwAA): clearly define task in the latter stages of the BPC process								X	X	X	X
CSwAA: hiding confidential information				X	X			X	X		
CSwAA: only crowdsource as a contest					X		X		X		X

Table 6.2 Decision table: people

Conditions			
The crowd for task: Available (A) vs. Not available (N)	N	A	A
Employees for task: Few (F) vs. Many (M)	–	F	M
Actions			
Not to crowdsource	X		
Should crowdsource		X	
CSwAA: consider other factors			X

acquisition of complementary resources and capabilities becomes necessary" (p. 5). Finally, if both conditions in Table 6.2 are satisfied, we suggest that crowdsourcing is still a good choice, yet advise further analysis of other factors, like task properties and management factors, before making the final decision.

Table 6.3 summarises the decision factors regarding management aspects: budget, availability of crowdsourcing experts, risk, and internal commitment. Several studies suggest that a sufficient budget is required for a crowdsourcing decision (Djelassi & Decoopman, 2013; Lofi, Selke, & Balke, 2012; Lu et al., 2015). Though the cost of crowdsourcing activities is usually small, other related costs such as quality control, service costs, coordination costs, and incentive mechanisms, may be significant. As a result, crowdsourcing should be appropriate

Table 6.3 Decision table: management

Conditions													
Budget: Sufficient (S) vs. Large (L)	S	S	S	S	S	L	L	L	L	L	L	L	L
Crowdsourcing expert: Available (A) vs. Not available (N)	N	A	A	A	A	A	A	A	A	N	N	N	N
Acceptance level of risk: High (H) vs. Low (L)	–	H	H	L	L	H	H	L	L	H	H	L	L
Internal commitment: High (H) vs. Low (L)	–	H	L	H	L	H	L	H	L	H	L	H	L
Actions													
Not to crowdsource	X												
Should crowdsource		X				X							
CSwAA: hire outside experts (due to large budget)										X	X	X	X
CSwAA: implement mechanisms for controlling risks				X	X			X	X			X	X
CSwAA: implement strategies for increasing internal commitment			X		X		X		X		X		X

for projects where the budget is sufficient. That is, the budget is not enough to perform the tasks in a traditional way, i.e. internal sourcing or outsourcing (Malone et al., 2010), but enough to cover the BPC process (Lu et al., 2015). In addition, crowdsourcing expertise and experience is necessary to coordinate the activities (Muhdi et al., 2011; Rouse, 2010). Thus, if a project has limited crowdsourcing expertise, hiring outside experts should be considered. Where hiring cannot be arranged due to limited budget, the project should not be crowdsourced.

Crowdsourcing also needs the project to have a high level of risk acceptance and internal commitment. As crowdsourcing relies on anonymous members of the crowd, it involves several risks, including low quality results and loss of intellectual property (Kannangara & Uguccioni, 2013; Naroditskiy et al., 2013; Schenk & Guittard, 2011). Consequently, like any other sourcing projects, mechanisms for controlling risks should be implemented. Furthermore, another factor that can jeopardise the adoption of crowdsourcing is to have a low level of employees' commitment to crowdsourcing (Brabham, 2008a; Lüttgens et al., 2014; Simula, 2013). This is because internal employees may fear losing their jobs because of crowdsourcing and thus create barriers for its adoption. To increase internal commitment, we suggest empowering key individuals who drive the crowdsourcing project (Lüttgens et al., 2014), and restructuring the internal incentive systems similar to what has been done in open innovation in order to overcome employees' negative attitudes (Huston & Sakkab, 2006).

Finally, as the lone environmental factor of the decision framework (Fig. 4.2), the availability of crowdsourcing platforms should be evaluated. Several researchers suggest the high availability of platforms is often critical for crowdsourcing activities (Chanal & Caron-Fasan, 2010; Lüttgens et al., 2014; Zogaj et al.,

2014), though it is also possible to build an organisational crowdsourcing platform. The reasons for adopting existing platforms include the large pool of crowd members (Mason & Suri, 2012), low setup efforts (Wang, Hoang, & Kan, 2013) and, in some cases, protection of intellectual property (Feller et al., 2012). Agreeing with these benefits, we note that there are two types of existing platforms: specialised and horizontal platforms. Specialised platforms concentrate on particular types of tasks, e.g. InnoCentive for problem solving tasks (Hirth et al., 2011), and thus have their own specialised members. Horizontal platforms, like AMT, may address different types of tasks and thus have diverse crowd members (Pavel Kucherbaev et al., 2013). This distinction may also influence the choice of using existing platforms. For instance, a crowdsourcing project having multiple dissimilar tasks may be more suited to a horizontal platform than a specialised one.

In summary, these decision tables have captured the reasoning rules for making the decision to crowdsource. With these reasoning rules, we are now ready to develop the decision tool.

6.2 Tool Development

Looking back to Fig. 6.1, the tool architecture was used to guide our development. As this development followed the rapid prototyping method (Lim et al., 2008), it was realised through two prototypes. Both prototypes were designed as web applications, using PHP and MySQL. The first prototype was developed and assessed in order to provide revision feedback. The second prototype development used this feedback to improve its functionality, and served as a tool supporting decision makers making informed decisions in BPC establishment. The following sections describe each of them.

6.2.1 The First Prototype

We developed a decision prototype to support the establishment of BPC. Concerning two main types of decision-makers in the BPC process, namely project managers and process designers, the prototype was designed with two main functions: Tool 1 and Tool 2. Tool 1 defines the project context, and analyses whether to crowdsource or not using the decision tables (Tables 6.1, 6.2, and 6.3). As a result, Tool 1 may suggest whether crowdsourcing is an appropriate choice for the project or not, and possible actions that may increase (or decrease) the probability of crowdsourcing. Tool 2 suggests the main workflows in a BPC process, i.e. task design, workflow design, crowd management, incentive mechanism, quality control, technical configuration and output (aligning with the innermost layer of the ontology—Fig. 5.2). Within each workflow, the tool suggests activities that should be operationalised and design options that support the operationalization. The outputs of Tool 2 are the concrete designs and related what-if advice. The GUI of the two tools are presented in Figs. 6.2 and 6.3.

Even though the two tools served different purposes, they were intentionally designed with a consistent user-interface. The GUIs were organised in four areas (in both Figs. 6.2 and 6.3). The left-hand side was dedicated to user inputs, allowing users to navigate within pre-defined issues (decision issues and design issues). These issues were aligned with the overview diagrams of the decision framework and design process, presented in the right-hand side. The middle area presented a pre-defined question and input parameters according to the chosen issue. When the user answers the question, the tool provided appropriate advice based on the reasoning knowledge.

This prototype had to be evaluated to provide feedback for the next round of development. To evaluate the prototype, a pilot experiment was conducted. The aim of the experiment was twofold. The first aim came from the prototyping perspective that determined whether the tool met its performance requirements and thus helped to identify possible improvements. The second aim, which originated from the evaluation point of view, sought to test the experimental materials for the evaluation of the tool (Dennis & Valacich, 2001). In the subsequent section, we focus on the first aim, while the second one will be discussed in the later sections.

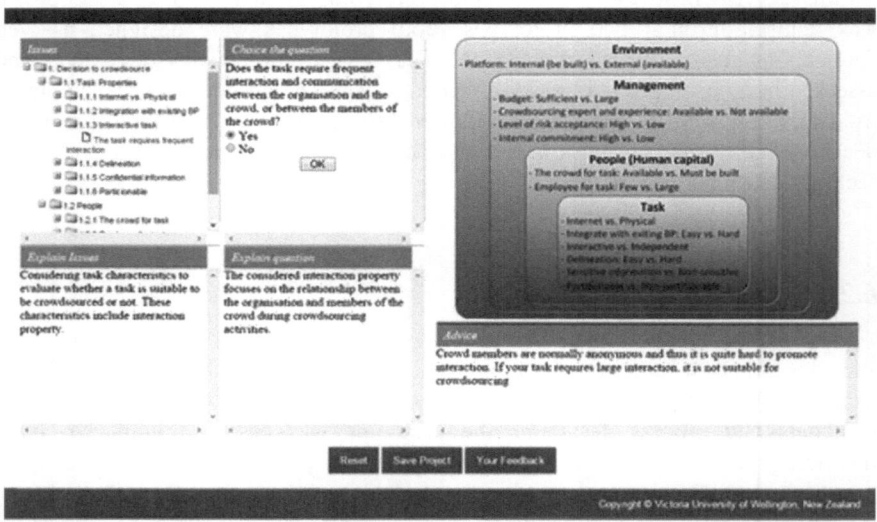

Fig. 6.2 Tool 1: The decision to crowdsource

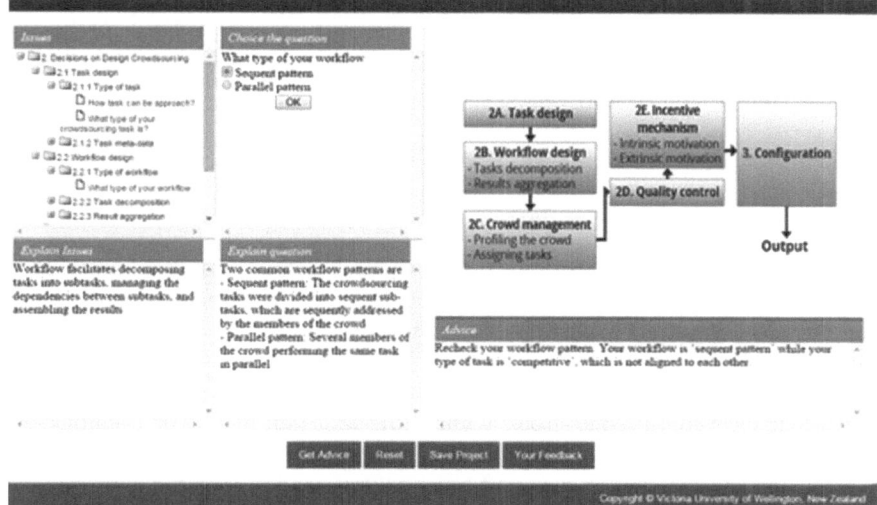

Fig. 6.3 Tool 2: Process design of BPC establishment

6.2.2 Pilot Experiment

The prototype was evaluated through a pilot experiment. In the experiment, we grouped participants into two groups: A and B. We asked all participants to perform two exercises, which were designed according to the decision to crowdsource and the process design. In exercise 1 (testing Tool 1), group A was the control group performing the exercise without the tool, while group B was the treatment group using the tool. In exercise 2 (testing Tool 2), the roles of the two groups were swapped to test Tool 1 and Tool 2 independently. The performance of participants was measured through their answers on the two exercises. In this experimental design, the prototype would be considered useful if the treatment group's performance outperforms the control group's performance. Besides analysing performance, we also observed how the participants used the tool during the experiment. This observation provided feedback to revise the prototype.

Overall, 49 students participated in the experiment. They were second and third-year IT students in the Can Tho University of Technology (CTUT). During the experiment, three of them did not use the prototype as requested, and thus their answers were removed from the final dataset. As a result, there were 46 participants remaining: 26 participants in group A and 20 participants in group B. Table 6.4 presents the final sample.

Pilot Experiment Results
We received 46 valid answers from the participants. Based on pre-defined standard answers, we calculated solution scores for each answer. More precisely, we used the

Table 6.4 Design of pilot experiment

	Exercise 1: decision to crowdsource	Exercise 2: crowdsourcing process design
Group A (26 participants)	*Control group* (without Tool 1)	Treatment group (using Tool 2)
Group B (20 participants)	Treatment group (using Tool 1)	*Control group* (without Tool 2)

following formulation: A correct answer was scored 1; 'No Idea' was scored 0.5; a wrong answer was scored 0. With the formulation, the participants' score on each exercise was calculated. Given that each exercise consisted of four questions, the score scale ranged from 0 to 4.

Figure 6.4 presents the score distribution in the experiment. Via Fig. 6.4, it seems that in exercise 1 the group without the tool (left-hand side columns)out-performs the other group (right-hand side columns), while in exercise 2 it is the group using the tool has higher performance (left-hand side columns). This can partly be seen via the highest scores of each group in both exercises (the last columns in Fig. 6.4). To further confirm this observation, we statistically analysed the data. As the data are not normally distributed, as suggested by the Shapiro-Wilk tests, we analysed the data using Mann-Whitney tests (Pfeiffer, Benbasat, & Rothlauf, 2014). The p-values of the Mann-Whitney tests are presented in Tables 6.5 and 6.6.

Tables 6.5 and 6.6 show mixed results regarding the usefulness of the tools. More precisely, the group using Tool 1 has a lower score compared to the other group (Table 6.5), while the group using Tool 2 has a higher score compared to the other group (Table 6.6). However, the differences between the groups with and without the tool in both exercises are not significantly supported at the significant level of 0.05 (p-value = 0.08 for exercise 1 and p-value = 0.51 for exercise 2). Given these results, we cannot reject the assumption that there are differences between the treatment and control groups.

Although the differences are not significantly supported, they indicate two important points. Regarding Table 6.6, the slightly higher performance of partici-pants using the tool (mean of 3.31 versus 3.18) is likely enough to warrant further study using Tool 2 for supporting the crowdsourcing process design. From the

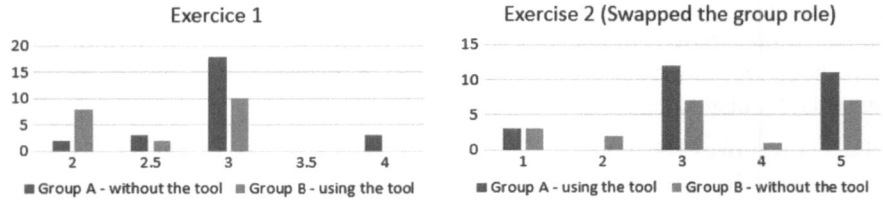

Fig. 6.4 Pilot experiment: score frequency in the two exercises

Table 6.5 Descriptive data and p-value: Exercise 1

Exercise 1	p-value	Without Tool 1 (Group A)			Using Tool 1 (Group B)		
		N	Mean	Std.	N	Mean	Std.
Solution score	0.08	26	2.98	0.48	20	2.55	0.48

Table 6.6 Descriptive data and p-value: Exercise 2

Exercise 2	p-value	Without Tool 2 (Group B)			Using Tool 2 (Group A)		
		N	Mean	Std.	N	Mean	Std.
Solution score	0.51	20	3.18	0.73	26	3.31	0.68

results, we expect that in further evaluation the tool will show its usefulness in designing the crowdsourcing processes. Regarding Table 6.5, participants using the tool are less performance than the others (mean of 2.55 versus 2.98). This suggests that this early prototype may not fully satisfy the design requirements. Thus, it is necessary to analyse participants' feedback for revision and development of the second prototype.

6.2.3   Feedback for Revision and the Second Prototype

Given the mixed results of the first prototype performance, this step assessed and identified feedback to further improve the tool. For this purpose, we analysed the participants' comments and our observation of how the participants interacted with the tool during the experiments. The analysis led to three important notes for the tool revision.

First, the tree structure of the prototype offered an effective way to access decision/design issues. We observed that participants preferred to use this structure to navigate to the issues that were most related to their context rather than step-by-step accessing them, which was aligned with what managers and process designers actually do in practice. As a side effect of too fast navigation, we observed that some participants jumped into the final decision (and answered questions in the exercises) based only on addressing one certain issue, rather than considering all related issues. This might lead to incorrect answers due to incomplete context awareness, which was particularly true for exercise 1, where the decision to crowdsource required considering all related factors. This might partly explain the unexpected results of using Tool 1. Given the discussion, one necessary revision was to keep the tree structure for navigation and to add a function for the final project's advice, complementary to the issues' advice.

Second, backing the tool with the BPC ontology was appropriate. During the experiment, the participants were effectively supported by having concept explanations, alternative parameters, and advice. This effectiveness could be clearly seen with Tool 2, which increased the performance mean in Table 6.6. Such support was enabled in the tool through the ontology. The appropriateness of using ontologies for founding the DSS design is suggested by Miah et al. (2014) and Amailef and Lu (2013) regarding the use of ontologies supporting DSSs in rural business operators and emergency response respectively.

Finally, besides our own observation, we also asked participants at the end of the experiment to give feedback for improving the tool. We received some major feedback related to loading time and working space. Regarding the loading time, some participants complained about the long waiting time. As the first prototype did not use interactive programming languages, it loaded a few times for presenting questions, parameters, question definition and advice. The loading time became longer when the number of participants accessing the tool increased. Addressing this issue, the next prototype should provide the information instantly, which required using interactive programming languages like JavaScript. Regarding working space, some participants suggested removing the model and process diagrams in the tools in order to provide more working space, i.e. issues, questions, and advice. Basically, the users preferred faster interaction and a simpler interface where the issue and advice functions are the focus.

The Second Prototype

Presenting in Figs. 6.5 and 6.6, the second prototyping addressed the three aforementioned notes. First, it continued using the tree structure for navigation, and further added the project's advice (area 1 in the figures). Consequently, users can receive both advice for a particular aspect and integrated advice for the whole project. Second, we kept using the BPC ontology for backing this prototype. Furthermore, the role of the ontology was extended to serve as a basic profile of crowdsourcing projects. This profile can be adapted regarding project conditions and intervention plans. Through this adaption, the tool can detect any inconsistencies in the input data, and provide advice for the whole project. Finally, the user interface of the second prototype was also updated.

As a result, the user-interface of the second prototype consists of three main areas: issue, input, and advice. The issue area is located on the left-hand side. It displays all decisional elements involved in crowdsourcing a business task, including the decision to crowdsource and the crowdsourcing process. The area is presented as a tree structure, reflecting the hierarchical structure of the BPC ontology. Users have two ways of navigation: (1) they may choose sequential navigation using the 'Next question' button and access each and every element; or (2) they may use the navigation tree to select and interact with specific elements of interest. With this navigation, the tool guides the users through the essential activities/decisions of the crowdsourcing process.

When the user selects an element, the input area is displayed on the right-hand side. This area presents a set of pre-defined questions that the user should answer. It

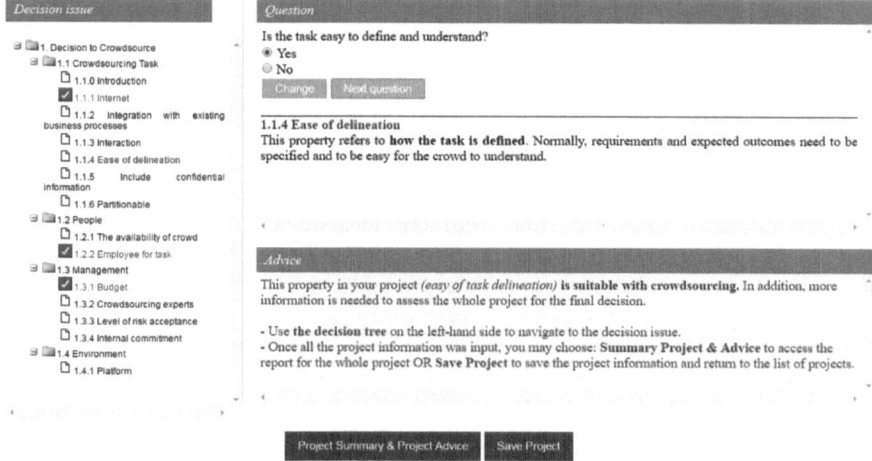

Fig. 6.5 Second prototyping: decision to crowdsource (Tool 1)

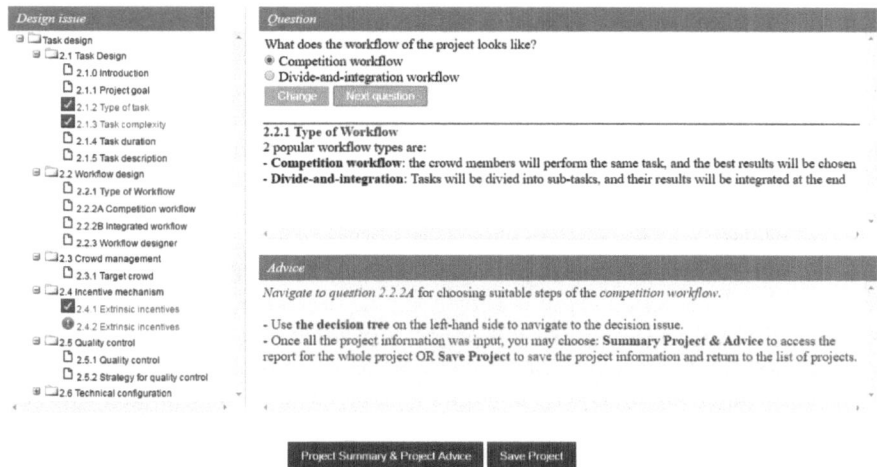

Fig. 6.6 Second prototype: process design (Tool 2)

also shows optional parameters that the user may select (presented as radio buttons or checkboxes), along with explanations about the questions and the parameters. The questions, parameters, and explanation reflect the semantics defined by the ontology. With this area, the tool offers understanding about the detailed elements of the crowdsourcing process. Further, the tool retrieves data to contextualise the crowdsourcing project, which will be used in the advice area.

The final area provides advice to the user. The tool combines the data input by the user with business rules and what-if scenarios defined by the BPC ontology, so that the advice provided is contextualised to the business task. Two kinds of advice are provided. First, the tool provides ad hoc advice, in relation with a specific input data. For instance, in Fig. 6.5 when the user declares that the task of a crowd-sourcing project is easy to define, the tool suggests the task is 'suitable with crowdsourcing' (The advice-box of Fig. 6.5). Second, the tool further assesses interdependency among the inputs and provides advice for the whole crowd-sourcing project. The user accesses this function through the button 'Project Summary & Project Advice', after she provides inputs for all elements of the crowdsourcing project (an example is provided below).

For better understanding of how the decision tool works, an example using tool 1 is presented. Company X has 10,000 pictures about wild animals that were captured by motion-triggered cameras in 40 wild locations in New Zealand. In this project, the main tasks are to identify the animals and their names in the pictures. These pictures are independent and thus the tasks can be performed individually. Company X wants help with finding whether crowdsourcing is appropriate for the project.

Accessing tool 1, the manager of company X sees a set of decision factors (left-hand side of Fig. 6.5). Reflecting the project, she chooses to interact with four factors and fulfils the answers for them: (1) Internet: the task can be done through the Internet; (2) easy of delineation: the task can be easy to define; (3) employee for task: the number of employees for the task is small; and (4) budget: the company is dedicated a large budget for the task. We note that the tool reminds the manager the two factors of employee for task and budget that she could forget if not using the tool. After filling in all information, the manager chooses 'Project Summary & Project Advice' to assess the whole project. The tool compares the provided answers with the rules captured by the BPC ontology. Given that the inputs of factor one, two, and three all influence positively on the decision to crowdsource (Afuah & Tucci, 2012), the tool suggests this project is *suitable with crowdsource*. Further, as the project has large budget (factor four), the tool also suggests *using this budget for hiring crowdsourcing experts to ensure the success of crowdsourcing*. Now, the manager can further explore different what-if scenarios by fulfilling answers for other factors and re-assessing the project until she reaches the most appropriate decision.

In what follows, we detail the empirical evaluation of the second prototype for its utility and perceived usefulness using experiments and focus groups.

6.3   Experiment

6.3.1   Experimental Overview

To collect empirical evidence about the utility of the decision tool, we conducted a series of six experiments. The use of experiments as an evaluation method has been suggested by design science literature (Hevner & Chatterjee, 2010; Peffers et al., 2012; Venable et al., 2016). The suggestion is clarified by Mettler et al. (2014) that "design experiments are an ideal technique for testing newly developed artefacts and for systematically deriving design improvements" (p. 224). In the current study, the experiments can be further characterised as ex-post and artificial evaluation. According to Venable (2012), the ex-post characteristic refers to the evaluation of an instantiated artefact, which is the decision tool in our research. The artificial characteristic is related to the controlled nature of the experiments, which assesses the tool in a lab setting.

We designed and conducted the experiments following the standard procedures for experimental research suggested by Montgomery (2012). The aim of the experiments was to study the utility of the tool in supporting the BPC decisions. For this purpose, we defined utility as 'having the ability to make difference on the performance between participants using the tool and others who do not use the tool'. Naturally, the experiments followed a simple comparative model with two conditions: using and not using the tool (Montgomery, 2012). This defined the independent variable of the experiments as the level of decision support: the usage and absence of the tool. Given the two supporting levels, groups of participants that used the tool were treatment groups, while the others were control groups. The dependent variable was the decision-making performance. All participants were asked to perform two exercises, designed according to the decision to crowdsource and the process design. The performance was measured through the participants' answers on the two exercises. Intuitively, the tool would be considered useful if the treatment groups had a higher performance compared to the control groups.

Given the discussion and the tool consisting of two distinct functions, supporting the decision to crowdsource (Tool 1) and crowdsourcing process design (Tool 2), two hypothesises were proposed:

Hypothesis 1 (H1): The usage of Tool 1 leads to better performance on making the decision to crowdsource.

Hypothesis 2 (H2): The usage of Tool 2 leads to better performance on designing crowdsourcing processes.

We expected that both hypotheses could hold. That is, the decision tool could increase user performance in both activities of BPC and thus demonstrate its utility.

6.3.2 Experimental Design

As comparative experiments, we could simply design the experiments with the treatment groups using both Tool 1 and Tool 2, and the control groups without either of them. However, such a design may generate a learning effect. That is, the participants, when they come to Tool 2, already have experience with using Tool 1, which may prevent our ability to evaluate the two tools separately. Given that, we decided to swap the group roles in the two exercises, i.e. group A was the control group in exercise 1 and then was changed to the treatment group in exercise 2; the swapping role of group B was vice versa. Table 6.7 represents this role swapping.

Participants

Participants were recruited from students of the Can Tho University of Technology (CTUT), Vietnam. To keep the participants homogeneous, the recruitment was based on class units, which were a combination of major and study year in CTUT. In particular, we recruited second and third year students, with a background in Information Technology (IT) and Industrial Management (IM). As a result, six experimental sessions were formed. In each session, the participants were randomly assigned to group A and group B. The numbers of participants in each session are presented in Table 6.8.

We acknowledged the concerns of using students as proxy for crowdsourcing practitioners, in particular, with respect to their crowdsourcing experience. However, the tool was designed in a way that general practitioners, with less experience, can benefit from using the tool. Thus, students with less experience are appropriate for testing the tool. Furthermore, the use of students as experimental

Table 6.7 Experimental design: swapping roles between the two groups

	Exercise 1: decision to crowdsource	Exercise 2: process design
Group A	Control group (without Tool 1)	Treatment group (using Tool 2)
Group B	Treatment group (using Tool 1)	Control group (without Tool 2)

Table 6.8 Numbers of participants per session (chronological order)

Session	Number of participants	Major	Study year	Group A (Exercise 1 without the tool; exercise 2 using the tool)	Group B (Exercise 1 using the tool; exercise 2 without the tool)
1	19	IM	Second	10	9
2	40	IT	Second	22	19
3	38	IM	Third	18	20
4	25	IT	Third	12	13
5	24	IT	Second	15	9
6	44	IM	Second	21	22

participants is very popular in software engineering (Sjøberg et al., 2005), and is considered appropriate to test computer-based tools (Dennis & Valacich, 2001).

Procedure and Materials

All sessions of the experiments were conducted between July and September 2015. Starting each session, participants were tutored to become familiar with the crowdsourcing concept and instructed on how to use the tools. They were then randomly placed into two computer labs according to groups A and B. Participants then had to complete exercise 1 and then exercise 2, each of which took about 30 min. The same exercises were delivered to both group A and group B. The only difference between the two groups was the treatment, where one group was instructed to use Tool 1/Tool 2 for addressing the exercises while the other group was not. At the end of each exercise, the participants handed their answers to the researcher.

The two exercises were developed in relation to the BPC decisions. In particular, exercise 1, focusing on the decision to crowdsource, included four different scenarios where crowdsourcing was a possibility. Each scenario had a short description and a question asking 'should the task [in the scenario] be crowdsourced?'. To make the scenarios diverse and close to practice, the scenario descriptions highlighted different decision factors, which were adapted from Afuah and Tucci (2012). Exercise 2 addressed issues related to the design of BPC processes. It consisted of two scenarios based on the case studies described in Sect. 4.3. Each scenario had a description and two design questions asking about different aspects of the BPC processes: task division, task description, incentive mechanism, and quality control. These aspects were adopted from Kittur et al. (2013).

Regarding the questions of the exercises, each of the two exercises consisted of four questions (yes/no and multiple-choice). We also asked the participants to explain their reason(s) for making the choice. The questions asked several crowdsourcing aspects, which were based on other studies to increase the neutrality of the experimental evaluation (Afuah & Tucci, 2012; Kittur et al., 2013). Also for the neutral purpose, when answering the questions, the participants might choose 'No Idea' if they thought that the scenarios did not provide enough information to answer a particular question.

By the end of experimental sessions, the participants were asked to complete a short survey on the perceived usefulness of the tool. The survey consisted of four questions adopted from Venkatesh and Davis (2000).

Pilot Experiment

A pilot experiment was conducted, serving two purposes. First, from a tool design point of view, the pilot experiment aimed at evaluating the tool for improving its design. Second, from an experimental point of view, the pilot experiment served as a test to refine the experiment materials (Dennis & Valacich, 2001). Since the first purpose and the results of the pilot experiment were already presented in Sect. 6.2.2 , the second purpose is presented here.

Learning from the pilot experiment, a few changes had been applied to the main experiments. First, the pilot experiment recruited only students with the IT background, which might lead to limited results in the decision to crowdsource since IT students might neglect a managerial focus. The main experiments extended the recruitment to include both IT and management students. The mix of students with different backgrounds also contributed to the robustness of the main experiment. Second, the lowest score in both exercises in the pilot experiments was two, which was quite high in a 0-to-4 range. This indicated that the pilot exercises were not complex enough to discriminate results (Dennis & Valacich, 2001). Addressing this issue, the main experiments increased the complexity of the exercises and asked participants to justify reasons for their answers. Finally, we already noted in the pilot experiment that some participants did not use the tool and answered the questions randomly. To increase the possible interaction, the main experiments asked participants to save their interaction with the tool.

Measurement

Solution scores were used to measure the participant performance. Due to the increased complexity of the exercises, scoring the participants' answers was not a straightforward task. Given that, a marking team was formed with four lecturers from CTUT (excluding the researcher). This team started by making standard answers and formulating the score. They formulated the score as: a wrong answer was scored 0; a correct answer was 0.5, plus meaningful explanation was an additional 0.5; 'No idea' was either 0 or 0.5 depending on the explanation. This meant each answer scored either 0 (zero), 0.5, or 1. Given the four questions and answers in each exercise, the score scale was from 0 to 4.

Using this formulation, the team started by marking together ten participants' answers and discussed any differences. After discussion and consensus on the marking, they did their marking individually. Additionally, the marking was arranged in a way that each session, i.e. both control and treatment groups, was marked by a single marker, which could reduce marking bias when comparing the score between the two groups. At the end of the marking process, the researcher compared the means of the scores among the sessions. One session had quite a high mean compared to others. A moderation meeting was organised to review the marks of the session, leading to a few changes on its scores.

6.3.3 Experimental Results

Overall, 190 students participated in the experiments. All participants' answers were scored and recorded in the sample. Table 6.9 shows descriptive statistics of the sample regarding each session. Starting the analysis, we tested the normality assumption of the sample. As the solution scores were treated as discontinuous measures (the minimum difference between two scores is 0.5), we expected that the normality assumption might not hold, which was confirmed by the Shapiro-Wilk tests, i.e. *p-values < 0.001 for both exercises*. These results strongly guided our choice of statistical tests in the following analysis.

Table 6.9 Descriptive results of six experimental sessions

Exercise	Session	N	Mean	Std.	95% Confidence interval for mean		Mean rank
					Lower bound	Upper bound	
Exercise 1	1	19	2.55	0.74	2.19	2.91	95.47
	2	40	2.40	0.95	2.09	2.70	89.10
	3	38	2.46	0.92	2.15	2.76	91.84
	4	25	2.46	0.83	2.11	2.80	90.78
	5	24	2.75	0.78	2.42	3.07	106.88
	6	44	2.61	0.92	2.33	2.89	100.97
	Total	190	2.53	0.88	2.40	2.65	
Exercise 2	1	19	2.71	0.89	2.28	3.13	98.74
	2	40	2.40	0.70	2.17	2.62	75.98
	3	38	2.71	0.84	2.43	2.98	98.84
	4	25	2.68	0.79	2.35	3.00	96.18
	5	24	2.96	0.72	2.65	3.26	113.29
	6	44	2.74	0.82	2.48	2.98	98.88
	Total	190	2.68	0.80	2.56	2.793	

We now look at the directions of measures within each session. Table 6.10 compares the measure means between the group using the tool and the group without the tool in the two exercises. Overall, the directions of measures were in line with our expectation in the two hypotheses. Almost all treatment groups had higher means than the control groups. One exception occurred in session 4 regarding exercise 1 where the treatment group had a lower mean than the other group. For all sessions (the last row of Table 6.10), the integrated means are consistent with the hypothesis directions.

Table 6.10 Comparison between groups using the tool and without the tool

	Exercise 1		Exercise 2 (Swapped the group role)	
	Group A—Without tool	Group B—Using the tool	Group A—Using the tool	Group B—Without tool
	Mean (Std.)	Mean (Std.)	Mean (Std.)	Mean (Std.)
Session 1	2.35 (0.85)	2.78 (0.57)	2.85 (0.88)	2.56 (0.92)
Session 2	2.34 (1.07)	2.47 (0.79)	2.71 (0.55)	2.03 (0.70)
Session 3	2.16 (0.84)	2.73 (0.92)	2.89 (0.78)	2.55 (0.89)
Session 4	2.54 (0.78)	2.39 (0.89)	3.00 (0.60)	2.39 (0.85)
Session 5	2.57 (0.75)	3.06 (0.77)	3.23 (0.68)	2.50 (0.56)
Session 6	2.48 (0.88)	2.75 (0.97)	3.25 (0.67)	2.22 (0.63)
All sessions	**2.40 (0.87)**	**2.67 (0.86)**	**2.99 (0.70)**	**2.34 (0.77)**

To use integrated data from these sessions, we first checked the potential differences of the scores among the sessions. As our sample datasets were not normally distributed, we used the non-parametric Kruskal–Wallis tests, which are an accepted alternative to ANOVA in case the datasets come from non-normally distributed population (Soh, Markus, & Goh, 2006). The results of the Kruskal-Wallis tests showed that there were no significant differences among the six sessions for both exercise 1 (*p-value = 0.788*) and exercise 2 (*p-value = 0.145*) at the 0.05 level. These results allowed us to analyse the datasets in an integrated way.

Using the integrated dataset, we tested the hypotheses H1 and H2. We chose Mann-Whitney tests to compare the different performance between the treatment and control groups because first, the tests were appropriate to the non-normally distributed population of the performance scores (Anderson, Sweeney, & Williams, 2011; Pfeiffer et al., 2014). Second, the discontinuous measures used in the study called for the use of non-parametric tests, which might lead to having higher power compared to parametric tests (Soh et al., 2006). Finally, the distribution-free nature of the Mann-Whitney tests placed few restrictions on the dataset, and thus allowed us to analyse the dataset integrated from six experimental sessions.

For each exercise, the Mann-Whitney tests were applied to the integrated dataset. We ran the tests using SPSS version 23.0. In this SPSS version, the software provided two ways to perform the Mann-Whitney tests: traditional (Legacy Dialogs) and new procedure (Nonparametric tests for independent sample). While the traditional procedure assumed the treatment and control samples had a similar-shape distribution, the new procedure actually tested this assumption. As the new procedure provided a more comprehensive test, we adopted and ran it on our integrated dataset. Regarding exercise 1, the results of the Mann-Whitney tests are presented in Table 6.11.

Regarding exercise 2, the results of the Mann-Whitney tests are presented in Table 6.12. We note the swapping roles of the two groups in exercise 2, group A using the tool and group B without the tool.

The experimental results shown in Tables 6.11 and 6.12 support the hypotheses H1 and H2. More precisely, the results show that the performance of the treatment groups were indeed higher than the control groups (mean rank = 104.27 vs. 87.44 regarding exercise 1, and 116.62 vs. 72.53 regarding exercise 2). Furthermore, the results confirm that the differences are significant at a 0.05 level in both exercise 1

Table 6.11 Results of Mann-Whitney tests on exercise 1

Exercise 1	p-value	Group A—Without tool				Group B—Using the tool			
		N	Mean	Std.	Mean rank	N	Mean	Std.	Mean rank
Solution score	0.03 (0.03)	99	2.40	0.87	87.44	91	2.67	0.86	104.27

Note The p-value of t-test is shown in parentheses for comparison purpose

Table 6.12 Results of Mann-Whitney tests on exercise 2

Exercise 2	p-value	Group A—Using the tool				Group B—Without tool			
		N	Mean	Std.	Mean rank	N	Mean	Std.	Mean rank
Solution score	<0.001 (<0.001)	99	2.99	0.70	116.62	91	2.34	0.07	72.53

Note The p-value of t-test is shown in parentheses for comparison

(p-value = 0.03 and U = 5,302.5) and exercise 2 (p-value < 0.001 and U = 2,414.0). From these results, we suggest to accept both hypotheses. In other words, the tool can help improve the participants' performance on BPC decisions. When comparing between the two p-values, we note that although the two hypotheses are both statistically supported, the support to accept H2 is stronger than H1, which will be further discussed in relation to the focus group evaluation. In summary, *we conclude that the two tools improve the BPC decision-making performance*.

Besides the performance analysis that showed the utility of the tool, we also analysed the tool usefulness perceived by the participants. We did this by analysing the survey data collected at the end of the experiments. The survey consisted of four questions rating the perceived usefulness (PU1–PU4). Of the 190 participants, 181 completed the survey. Table 6.13 shows the statistics of the survey results.

The results show that all items display a tendency towards perceived usefulness, i.e. all means > 3.90 on the 1-to-5 scale that varied from extreme uselessness (1) to extreme usefulness (5). In other words, the participants perceive the tool to be useful for their tasks. However, we note that the perceived usefulness here needs to be interpreted carefully. This is because when we examine the dataset of participants who did not perform well when using the tool (scoring less than or equal to 2 on the 0-to-4 scale), their perceived usefulness is still high (mean of 4.07). This indicates that the participants might answer the survey without considering the tool performance. An explanation for this is that the participants know that the researcher is also a lecturer in the university, leading to a potential bias when students evaluate their instructors (Marsh, 2007).

Table 6.13 Statistics of usefulness perceived by experimental participants

Perceived usefulness	Mean	Std.
PU1—Using the tool allows me to better answer the questions in the exercises	3.98	0.632
PU2—Using the tool allows me to faster answer the questions in the exercises	4.01	0.796
PU3—Using the tool allows me to better understand the questions in the exercises	3.96	0.766
PU4—I find using the tool useful	4.28	0.667

Note We used 1–5 scale to rate the PUs where 1 is useless and 5 is useful

Table 6.14 Validity of the experiments

Conclusion validity: concerns issues that may influence the correctness of the conclusion. Two aspects were focused: the use of appropriate statistical tests and avoiding irrelevancies in the experimental setting - We screened data for their appropriateness with the statistical tests. As the normality assumption did not hold, we used Mann-Whitney tests to analyse the difference between the control and treatment groups - We further analysed the possible influence of other factors like participants' background, gender, and year of study on the results. Both the chi-square tests and Mann-Whitney tests rejected hypotheses that the performance was different across different categories of background, gender, and year of study. Consequently, these results reduced the threads of irrelevant factors influencing the results
Internal validity: requires causal relationship between treatment and outcome - We randomly assigned participants to the control and treatment groups - To avoid learning effect, we swapped group roles in exercise 2 (the control group became the treatment group, and vice versa) - This validity also related to the dropout rate. In the pilot experiments, some participants did not use the tool to address tasks, leading to a small rate of dropout in the treatment group. The main experiments handled this issue by asking the participants to save their answers, which increased their interaction with the tool and thus minimised the dropout rate
Construct validity: concerns how the measures represent their theoretical basic - To measure the performance, we used two exercises consisting of four questions, which were developed based on previous studies (Afuah & Tucci, 2012; Kittur et al., 2013). Yet, we noted the exploratory nature of our study, which suggests that the constructs still need to be further analysed and tested
External validity: concerns the generalizability level of the results - Kruskal-Wallis tests were particularly motivated by external validity considerations. The tests confirmed that there were no significant differences between the six sessions, thereby increasing the external validity - As noted earlier, using students as proxies for crowdsourcing practitioners might threat an external validity. However, as the tools proved to be helpful for students with less crowdsourcing experience, they would certainly be helpful for crowdsourcing practitioners. Furthermore, Höst et al. (2000), who compared between students and professionals as experimental subjects in software engineering, found very minor differences between the two groups

Validity

To strengthen the knowledge claims from the experiments, we had to identify and handle several threats to validity (Wohlin et al., 2012). Table 6.14 presents those that were most relevant to the experiments, and summarises how we handled them.

In summary, we conducted experiments to evaluate the decision tool supporting BPC establishment. More precisely, we examined how much the tool helped making informed decisions for the scenario applications. We designed the experiments where the treatment groups used the tool and the control groups did not use the tool to address the same exercises. Two hypotheses were tested on using the tool to support the decision to crowdsource (H1) and the process design (H2). Six experimental sessions with 190 participants were conducted, which together formed an integrated dataset of the experiments. Since the dataset was not normally distributed, we used non-parametric tests to analyse the data. The results provide support for both hypotheses, which suggests that the use of the tool (comprising of Tool 1 and Tool 2) leads to better performance on BPC decisions. In short, the tool is useful in supporting BPC establishment.

6.4   Focus Group

6.4.1   Overview of the Approach

The previous section has already evaluated the tool through controlled experiments, which can have precision, but are not very strong in exploring the participants' perception. To examine the tool usefulness perceived by the participants, this section presents the focus group approach to evaluate the tool. We highlight how complementary the focus groups are to the above experiments, by examining two aspects. First, focus groups provide qualitative evaluation (Krueger & Casey, 2014; Tremblay, Hevner, & Berndt, 2012), complementary to the quantitative data of the experiments. Second, focus groups, having the strength of group discussion, can give us key interaction-based insights that may not surface in the experiments.

Adopting the focus group approach, our aim was to gather qualitative evidence of the tool's utility. This aim identified the nature of the focus group. Tremblay et al. (2010) categorised focus groups being used in design science into exploratory ones that generated design features of artefacts, and confirmatory ones that gathered evidences of the artefact utility. Considering our aim, the confirmatory focus groups were adopted. More precisely, we used the focus groups to confirm the tool utility perceived by the participants, regarding three aspects: (1) perceived usefulness that measures 'what ways the tool can contribute to BPC establishment'; (2) perceived ease of use that measures 'the ability that the users can comprehend the tools to perform their targeted tasks'; and (3) possible improvements of the tool.

6.4.2   Focus Group Design

To guide the focus group design, we adapted the procedure of how to conduct focus groups in design science proposed by Tremblay et al. (2010). Figure 6.7 presents a summary of the adapted procedure, consisting of problem formulation, sample frame and moderator, question route development, conduct of focus group, and data analysis. Besides the problem formulation that was already presented in the previous section, the rest of the procedure is presented in this section.

Sample Frame and Moderator
Four key decisions were made in this stage: the number of focus groups, the desired number of participants in each group, recruitment participants, and moderator identification. The literature was not clear about the number of groups necessary to

evaluate IS artefacts. For instance, Gibson and Arnott (2007) used one focus group to evaluate a business intelligence system, while Miah et al. (2009) conducted three focus groups to acquire knowledge backing the design artefact. More recently, Tremblay et al. (2010) suggested that focus groups for design science should include one pilot, two exploratory, and two confirmatory groups. Considering the suggestion, combining with the confirmatory nature of our evaluation, the current study conducted two focus groups.

Regarding group size, focus groups normally include a range from four to twelve participants. Within this range, Tremblay et al. (2010) explain some trade-offs between smaller and larger group sizes. For instance, smaller groups require group members to participate more, while larger groups increase complexity. These authors explicitly suggest using about six participants as "large focus groups (more than six) could be tricky in design research since the subject matter is more complex than traditional focus group topics" (Tremblay et al., 2010, p. 603). Following this suggestion, we recruited one group with four participants and the other group with six participants.

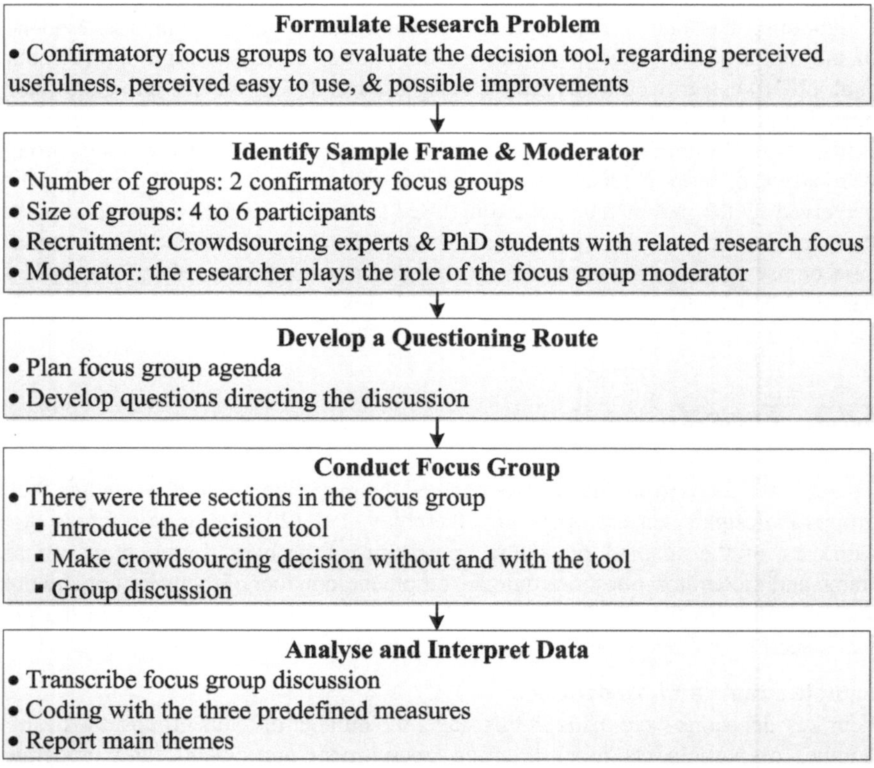

> **Formulate Research Problem**
> • Confirmatory focus groups to evaluate the decision tool, regarding perceived usefulness, perceived easy to use, & possible improvements

> **Identify Sample Frame & Moderator**
> • Number of groups: 2 confirmatory focus groups
> • Size of groups: 4 to 6 participants
> • Recruitment: Crowdsourcing experts & PhD students with related research focus
> • Moderator: the researcher plays the role of the focus group moderator

> **Develop a Questioning Route**
> • Plan focus group agenda
> • Develop questions directing the discussion

> **Conduct Focus Group**
> • There were three sections in the focus group
>   ▪ Introduce the decision tool
>   ▪ Make crowdsourcing decision without and with the tool
>   ▪ Group discussion

> **Analyse and Interpret Data**
> • Transcribe focus group discussion
> • Coding with the three predefined measures
> • Report main themes

Fig. 6.7 Focus group procedure (adapted from Tremblay et al., 2010)

Regarding the nature of the two groups, we recruited one with diverse crowd-sourcing backgrounds for assessing different perspectives of the tool, and one with homogeneous crowdsourcing experts for thorough assessing the tool. In the first group, we recruited four Ph.D. students with backgrounds covering crowdsourcing, social media, social network, and IS quality evaluation. This group helped assess the tool from the view of general users, who in practice might come to the tool with less crowdsourcing background. Furthermore, it served as a pilot test on the script, setting, and question route of the focus groups. The second group recruited six crowdsourcing practitioners and researchers, who had more than one year of crowdsourcing experience. The aim of this focus group was to thoroughly evaluate the tool from the view of crowdsourcing experts. Tables 6.15 and 6.16 show demographic characteristics of the focus group samples.

As focus groups need to be moderated, the researcher acted as a moderator of the focus groups. This role included introducing the decision tool to the participants, facilitating the discussion, and dealing with the dynamic in the group discussion. As the research was also the artefact designer, the researcher came to the moderator role with an open mind regarding the tool evaluation. That is, the researcher viewed the focus group as a good opportunity to receive suggestions for improvements, constructive feedback and (sometimes) criticism. During the focus group, the moderator sometimes answered questions on how to use the tool, as participants could not be expected to be completely familiar with the tool functionality after just a short introduction.

Questioning Route and Crowdsourcing Scenario Development

Before the focus groups, we developed a questioning route, which would set the direction of the group discussion. The questioning route included ten questions used by the moderator to initiate the discussion. Due to the pilot role of the first focus group, these questions were slightly revised in the second focus group.

Focus Group Conduct

The focus groups were held in a meeting room at the School of Information Management, Victoria University of Wellington. The room had several laptops

Table 6.15 Sample of focus group 1: Ph.D. students

Gender	Age	Crowdsourcing expertise	Current position/ occupation	Years of crowdsourcing experience	Research focus
Female	25–34	Researcher	Ph.D. student	<6 month	Crowdsourcing
Male	25–34	Interested in crowdsourcing	Ph.D. student	<6 month	Social network
Male	25–34	Interested in crowdsourcing	Ph.D. student	<6 month	IS service evaluation
Female	25–34	Interested in crowdsourcing	Ph.D. student Lecturer	<6 month	Social media

Table 6.16 Sample of focus group 2: crowdsourcing experts

Gender	Age	Crowdsourcing expertise	Current position/occupation	Years of crowdsourcing experience	Years of work experience
Male	45–54	Practitioner/ Researcher	Associate director library technology service	>2 years	>5 years
Female	>55	Practitioner	Contractor	>2 years	>5 years
Male	25–34	Researcher		1–2 years	1–2 years
Male	>55	Researcher	Senior Lecturer	>2 years	>5 years
Male	35–44	Practitioner	Platform owner Digital collections	>2 years	>5 years
Female	>55	Researcher	Lecturer	1–2 years	>5 years

accessing the decision tool. After the welcome, the moderator introduced the tool functionality to help the participants become familiar with the tool. Then, the moderator delivered two scenarios to the participants, each of which included a short description and five decisions related to the decision to crowdsource and process design. These scenarios, which were developed based on the two case studies in Sect. 4.3, ensured the participants used the tool and thus enabled them to discuss the tool. The moderator encouraged the participants played the role of a crowdsourcing decision maker addressing these scenarios. At the beginning, the participants were asked to make the decisions on the related BPC aspects without the tool. Then, they were asked to access the decision tool and reconsider their decisions using the tool. They were prompted to write out both their initial decisions and revised decisions (if there were any).

The ensuing discussion revolved around how the decision tool was used and its ability to support the decision-making process. Starting the discussion, the moderator asked some initial questions about the tool characteristics, and then allowed the discussion to flow. The discussion was audio recorded and transcribed. At the end of the discussion, both the initial and revised decisions of the participants were handed to the researcher. Some notes were also taken by the moderator during the discussion. All these activities took about 1.5 h for each focus group.

Analysis

For data analysis of design science focus groups, Tremblay et al. (2010) note several qualitative analysis techniques that can be used and highlight the use of template analysis due to its flexible and simple procedure. Template analysis is referred to as "thematic analysis that balances a relatively high degree of structure in the process of analysing textual data with the flexibility to adapt it to a need of a particular study" (King, 2012, p. 426). Following Tremblay et al.'s (2010) suggestion, we adopted the template analysis and further viewed the technique as appropriate for two reasons. Our focus groups evaluated the tool with three pre-defined measures: perceived usefulness, ease of use, and possible improvements, while one distinctive characteristic of template analysis is the use of pre-defined themes (Brooks, McCluskey, Turley, & King, 2015). Furthermore, this analysis technique can be used within the design science paradigm, as suggested by Tremblay et al. (2012).

Adopting the template analysis, we developed the initial templates based on the three measures and some lower codes focusing on some aspects of the measures. A code is defined as "a label attached to a section of text to index it as relating to a theme or issue in the data which the researcher has identified as important to his or her interpretation" (King, 2004, p. 257). We then applied the codes to the transcribed focus group discussions. More precisely, we reviewed the transcripts and identified sections of the text relevant to our codes. During this process, we created some other codes to explore the entire range of the discussion. Table 6.17 presents an example of the coding schema. The codes were then aggregated into categories and themes.

6.4.3 Focus Group Results

The results of the focus groups are structured according to the three investigated measures: perceived usefulness, perceived ease of use, and possible improvements. Regarding the first, perceived usefulness was further analysed with two aspects: the ability of the tool to provide additional and structured information to the participants, and the ability to change the participants' decisions after using the tool. For this analysis, the two codes 'additional information' and 'Decision framing' were examined. In general, there were mixed results regarding the two aspects. On the one hand, we found that the decision tool provided additional structured information for making the decisions, as demonstrated by the following comments:

Table 6.17 Example of coding schema

Measure	Code	Definition
Perceived usefulness	Decision framing	The ability to frame crowdsourcing decisions, which may lead to change in these decisions prior and after using the tool
	Additional information	The complementary information provided by the tool, in addition to what the participants had already known without the tool
Perceived ease of use	Ease of use	Participants can comprehend the tools to perform their targeted tasks
Possible improvement	Knowledge improvement	Suggestion to improve the tool regarding crowdsourcing information, factors, advice, and decisions
	Technical improvement	Suggestion to improve the tool regarding user interface and technical functions

It is definitely promoting a lot of the right things to help make a correct decision.

The tool makes it clear why it was saying not to do [crowdsourcing], so you feel confident, it was an informed decision.

The tool provides more concrete [information]. I have some abstract ideas, it helps the actually specifics.

Further analysing this aspect, we found quite a common template that the tool reminded the participants of something that they forgot, which highlighted the ability of the tool for providing structured information related to BPC.

It forces me to think about the risk which I haven't thought about when I did it [crowdsourcing].

With the tool, it brought up some privacy issues.

I think for me, this is absolutely helpful to say have you thought about this.

On the other hand, the supporting evidence for framing decisions was not strong. In other words, the results were mixed regarding how the tool could frame crowdsourcing decisions and change the participants' decisions. Some participants changed at least one decision as a result of using the tool. Reflecting that, when being asked whether they changed their decision after using the tool, some participants wrote out:

Yes, quality aspects have changed, which is good to point out.

Yes, the decision to crowdsourcing function [of the tool] influenced me.

Yes, [I] have added the role of the internal experts and have stated that this is a complex project.

However, other participants did not change their decisions. In some cases, the participants' thought covered the tool's framing, as seen via "the tool provides advice that follows my own understanding". In a few cases, even though the tool suggested different decisions, the users still kept their own decisions. As one of them stated that "I still rely on my own decision making".

Regarding perceived ease of use, most participants made very similar comments and suggested that the tool was clear and understandable. For example, one of them commented: "Yes, the tool is easy to use and I don't have any difficulty to learn it". This aspect was also confirmed by the fact that all the participants learned how to use the tool through a short introduction in the focus groups. They mastered the tool quickly, except one participant who needed further explanation about the tool's navigation during the focus group.

Finally, we also analysed the possible improvements for the tool, including knowledge improvements and technical improvements. The focus groups suggested a few knowledge improvements. Some participants suggested adding more decision factors and design issues that should be considered in the tools, including confidentiality, sustainability, timeline, life cycle, whether tasks can be automatic, and crowd engagement. For instance, one participant recommended that the decision to crowdsource should examine the confidentiality of the crowd. We note that, on the one hand, these suggested factors and issues would be interesting to explore further with the possibility of the tool revisions. On the other hand, as these factors and issues were not suggested by the 'wisdom of researchers' in our knowledge sources analysis (Sect. 3.1), they might not be key factors for different crowdsourcing contexts. Thus, the suggested factors and issues should be further examined before possibly including or excluding them into our tool.

Another interesting suggestion was to give different weights to the decision factors and issues, and ultimately to use these weights for aggregating and generating the final decision. One participant stated "the idea of weighting is interesting too, where some of these issues are perhaps more important than others". Although such a weighting approach was not included in the tool yet, it was aligned with the nature of the tool. That is, the tool was developed based on the 'wisdom of researchers' and thus the numbers of the supporting sources could be used as weights.

During the discussion, some technical improvements were also suggested. We noted that some suggestions were not actual problems after some explanation and discussion. For instance, a participant commented on why the tool did not immediately move to the next question after the user answered a question. The explanation was that after the user answered a particular question, the tool needed to show advice for the current question, and thus needed not to move to the next question. Besides that, other suggestions might improve the tool, and thus they should be further considered. Examples included a confusion where two issues in the design tool had the same headings; and some questions, pre-defined answers, and advice needed to be clarified, e.g. to provide more examples to clarify the questions and pre-defined answers.

Overall, the focus group results were positive towards the tool utility. Table 6.18 summarises the main evaluation findings of the focus groups. As seen in Table 6.18, we find clear evidence of perceived usefulness that the tool can provide additional structured information related to BPC decisions. It is also evidence of perceived ease of use. Yet, evidence of decision framing that the tool can change participants' decisions is mixed. Further, some possible improvements for the tool are also noted. In summary, we suggest that the tool provides promising support to BPC decisions though some improvements are desired.

Before concluding the section, we recall a note from the experiment results. That is, the experimental results supported the utility of both main functions of the tool (Tool 1 and Tool 2), yet the statistical support of Tool 2 was stronger than Tool 1. This note can now be explained with the focus group results. To clarify, Tool 1 focuses on the decision to crowdsource, which is a go/no-go decision. The focus group results show that this type of decision is quite hard to change (mixed results on decision framing), which is aligned with the equal support of the experimental results for the Tool 1's utility. Differently, Tool 2 focuses on the BPC design issues. The focus group results show that providing structured information is largely helpful for these issues (strong results that the tool can provide structured information). This possibly explains why the experimental results strongly supports the Tool 2's utility.

Table 6.18 Summary of focus group results

Focus group	Supporting evidence	Counter-evidence	Evaluation results (Evidence of utility)
Perceived usefulness—*Providing additional information*			
FG 1	Several instances where the tool was useful to provide information and structures for the BPC decisions	None	**Yes**
FG 2	A common template where the tool reminded the participants something that they did not think of without the tool	None	**Yes**
Perceived usefulness—*Decision framing*			
FG 1	After using the tool to frame crowdsourcing decisions, some participants changed at least one decision	A few participants made a decision on their own knowledge	**Mixed**
FG 2	After using the tool to frame crowdsourcing decisions, some participants changed at least one decision	Some participants suggested the tool did not change their decisions	**Mixed**
Perceived easy to use			
FG 1	Most participants suggested the tool was easy to understand	None	**Yes**
FG 2	Most participants suggested the tool was easy to understand	One participant needed support on how to use the tool	**Yes**
Possible improvement—*Knowledge improvement*			
FG 1		A few suggestions for knowledge improvements	**High** (a few suggested improvements)
FG 2		Some suggestions for additional decision factors/ design issues, and for weighting them	**Average** (some suggested improvements)
Possible improvement—*Technical improvement*			
FG 1		**Some suggestions** for technical improvements	**Average** (some suggested improvements)
FG 2		**Some suggestions** for technical improvements	**Average** (some suggested improvements)

In conclusion, the focus group results provide qualitative evaluation on the tool, complementary to the experimental evaluation. The focus groups generate rich discussion and assessment the tool from different angles. These rich data help point out what areas the tool can make contributions and what still need to be improved. As a result, the qualitative evaluation here, together with the quantitative evaluation from the experiments, strengthens the confidence on the utility of the decision tool.

6.5   Summary and Discussion

This chapter constructed a decision tool supporting BPC establishment, operationalising the ontological knowledge base developed in the previous stage. Considering the complexity of the construction, we developed a tool architecture and adopted a rapid prototyping method (Kordon, 2002; Lim et al., 2008). The tool architecture consisted of three main modules: GUI, information-processing module, and a knowledge module that was backed by the BPC ontology (Chap. 5). We used the architecture to develop two prototypes. While the first prototype enabled us to test the tool for revision feedback, the second one was targeted to support the project managers and process designers, making informed decisions in BPC establishment.

The decision tool was carefully evaluated. We assessed the tool using experiments as quantitative evaluation and focus groups as qualitative evaluation. In the experiments, 190 participants were asked to address two crowdsourcing exercises. They were allocated to the control groups without the tool and the treatment groups with the tool. The results find significantly higher performance of the treatment groups compared to the control groups in both exercises. This suggests the tool is useful in supporting BPC decisions.

To further evaluate the tool, two focus groups were conducted. One group consisted of crowdsourcing experts, and the other consisted of Ph.D. students with the related backgrounds. In each focus group, the participants considered some BPC decisions without and then with the tool. Based on the interaction with the tool, they then discussed about the tool functionality and its support for making BPC decisions. The focus group results likely confirm the utility of the tool, especially regarding its ability to structure and provide useful information in establishing BPC.

The combination of the two evaluation techniques increases our confidence when suggesting the tool as a means for supporting BPC establishment.

Chapter 7
Discussion and Conclusion

> *The ultimate assessment for any research is 'What are the new and interesting contributions?'.*
>
> —Hevner et al. (2004)

The preceding chapters of the book have presented the detailed findings of the research. This chapter examines them from a more integrated perspective, highlighting the interrelated nature of the research and positioning the research contributions. The chapter starts with the interrelated findings and our consolidation of the findings for addressing the research objectives. We then discuss four major contributions of the research. Following this is a discussion of contributions towards organisational practice. Then, limitations are discussed. Finally, we conclude the book and outline future research.

7.1 Interrelated Results

The research results have been formed from the four research stages. While the previous chapter presented the results as sequential stages' outcomes, these results are related due to the interrelated nature of the research. This section examines the results from an integrated viewpoint in order to provide an overall picture of the research outcomes. In particular, four major integrated outcomes are discussed.

First, we note that *the research results are interrelated in structuring the BPC domain.* This is because the four research stages together examine the BPC domain, aligning to the exploratory-confirmatory continuum suggested by Miles et al. (2014). Figure 7.1 illustrates this interrelation. The first stage explored the knowledge sources in the domain, which had not been structured before. The second stage deducted the knowledge sources and conceptualised the BPC concept. It offered a conceptual model that synthesised unstructured knowledge into the focused building blocks of BPC. The third stage, extending this conceptualisation, organised knowledge in the domain using an ontological structure. The final stage instantiated a decision tool founding on the ontological structure, and then

© Springer International Publishing AG, part of Springer Nature 2019
N. H. Thuan, *Business Process Crowdsourcing*, Progress in IS,
https://doi.org/10.1007/978-3-319-91391-9_7

evaluated and confirmed the tool utility. In reflecting through the research stages, BPC knowledge has been sequentially structured, starting from diverse unstructured knowledge sources, to abstract conceptualisation, to an ontological structure, and finally to the instantiated decision tool supporting BPC establishment. Consequently, we suggest that the research results enable different *yet interrelated knowledge for structuring BPC*.

Second, the research results also suggest interrelated *yet different levels of abstraction for understanding BPC establishment*. This difference allows us to speak both abstractly about managerial aspects of BPC, and more concretely about its building blocks and detailed processes. The conceptual model, ontology, and decision support tool form three levels of BPC abstraction, which are depicted in Fig. 7.2. In the figure, the conceptual model presents abstract building blocks of BPC; the ontology specifies these building blocks into detailed elements, including processes, activities, data, and their relationships; and the decision tool operationalises these ontological elements with decision tables, what-if scenarios, and contextual recommendations. Given the three levels of abstraction, it is possible for different stakeholders to focus on different levels of concern but still reach consistency on BPC establishment. These consistent yet different levels of foci are an important requirement to establish complex business processes involving multiple stakeholders like BPC (Berente et al., 2009; Giachetti, 2004; Hasselbring, 2000).

The first two outcomes lead to the third interrelated result, which is *the ability to trace back the BPC knowledge through the research stages* (the upward arrow of Fig. 7.2). That is, operationalised knowledge in the decision tool can properly be traced back to the ontological elements, which can be mapped to the model components and in turn traced back to the knowledge sources. The traceability comes from the systematic approach brought by the design science research, where we systematically structure the research activities and explicitly justify and present key decisions made in these activities. This systematic approach is similar to the evidence-based strategy in design science (Denyer & Tranfield, 2006; Van Aken, 2005; Van Aken & Romme, 2012).

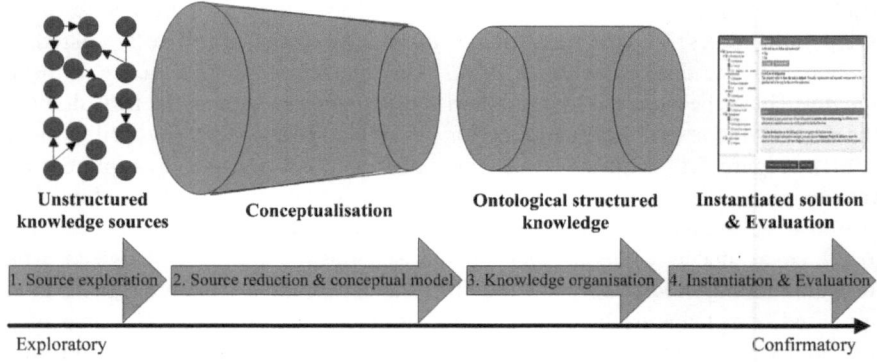

Fig. 7.1 Interrelated knowledge in structuring BPC

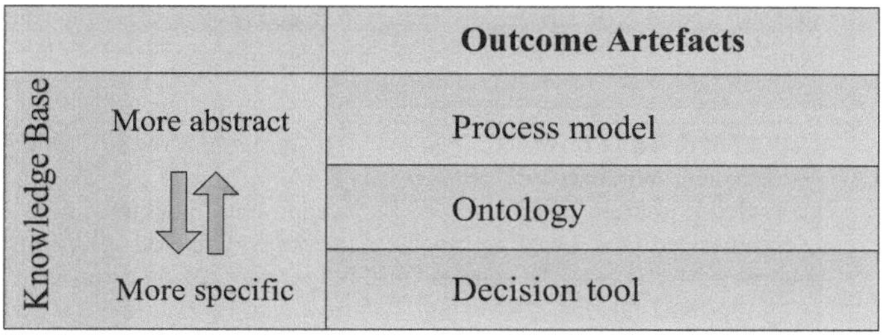

		Outcome Artefacts
Knowledge Base	More abstract	Process model
	⇕	Ontology
	More specific	Decision tool

Fig. 7.2 Interrelated yet different levels of abstraction: outcome artefacts

The fourth and final integrated outcome is *the multiple iterations of build and evaluate activities* in the research process. Inspired by the design cycle suggested by Hevner and Chatterjee (2010), we designed and then evaluated each artefact before moving to the next research stage. These iterations allow us to better understand the design problem, its solution, and how the solution addresses the problem through evaluation. The iterations also enhance the relevance and rigour of the research stages by continuous evaluating the outcome artefacts (Hevner & Chatterjee, 2010; Sonnenberg & vom Brocke, 2012b). As a result, the build-evaluate iterations strengthen the relevance and rigour of the entire research process and its generated BPC knowledge.

7.1.1   Addressing the Research Objectives

This section summarises the research results in order to address the research objectives. Four research objectives have guided the book, which are recollected here.

1. RO1: To understand the main building blocks of BPC that can be identified in the domain.
2. RO2: To develop a model structuring the identified building blocks for conceptualising BPC.
3. RO3: To construct a domain ontology of BPC that organises the unstructured knowledge sources in the domain.
4. RO4: To construct a decision tool supporting organisations in establishing BPC.

These research objectives have been realised explicitly in Chaps. 3, 4, 5, and 6 respectively, which are now summarised. Table 7.1 provides a structured summary of main results that address each research objective. The table is organised around four columns. The first column shows the four stages of the research (presented in Chaps. 3, 4, 5, and 6). Then, as design science highlights both design processes and design products (Hevner & Chatterjee, 2010), the second and third columns present the research activities and research outcomes respectively. The final column refers to the research objectives being addressed in each research stage.

Table 7.1 Summary of results that answer the research objectives

Research stage	Research activity	Research outcome	RO
1. Scoping knowledge sources	**BPC Knowledge Base** • Identified and analysed 238 knowledge sources • Synthesised BPC building blocks supported by at least 10 sources • Synthesised factors influencing the decision to crowdsource	• 12 building blocks of BPC (Table 3.2) • Additional outcomes: nine factors and sixteen sub-factors influencing the decision to crowdsource (Table 3.3)	RO1
2. Develop the IS Artefact	**Conceptual Model** • Synthesised the most salient BPC building blocks • Applied the analytic framework to arrange the model components • Defined the components • Developed a framework supporting the decision to crowdsource	• A process model of BPC (Fig. 4.1) - 3 stages: decision to crowdsource, design, and configuration - 7 components in the three stages • Additional outcomes: A decision framework of the decision to crowdsource (Fig. 4.2)	RO2
	Case Study Evaluation (two crowdsourcing projects) • Collected multiple data sources, including interviews with key informants • Analysed the project activities, using the model • Analysed the utility of the model perceived by the interviewees	High representation of the projects' activities (Figs. 4.3 and 4.4) • Usefulness perceived by the interviewees for planning and running crowdsourcing projects (Sect. 4.3.3)	
3. Develop the IS Artefact	**Domain Ontology of BPC** • Ontology capture - Analysed the knowledge sources in detail - Identified ontological elements: concepts, hierarchical relationships, and decision-making relationships • Knowledge organisation - Synthesised the ontological elements - Organised the ontological elements using a layered structure	• Lightweight ontology of BPC (Fig. 5.2) - 39 salient concepts (Table 5.1) - Five types of hierarchical relationships • Heavyweight ontology of BPC - Decision-making relationships (Sect. 5.2.3), which turns the lightweight into the heavyweight ontology	RO2, RO3
	Triangulation Evaluation • Compared the BPC ontology with a version generated by OntoGen - Took abstracts of the same knowledge sources as input - Used OntoGen to generate an ontological version	• High coverage of domain concepts and relationships • High clarity of the domain semantics • Our ontology provides clearer meaning and capturing both hierarchical and decision-making relationships.	

(continued)

Table 7.1 (continued)

Research stage	Research activity	Research outcome	RO
	- Compared our ontology with the generated version		
4. Develop the Instantiated Artefact	**Decision Tool** • Based on the ontology • Developed two prototypes - Used the first one for gathering feedback - Developed the second prototype based on the feedback	• A decision tool with two main functions - Supporting the decision to crowdsource (Fig. 6.5) - Supporting process design (Fig. 6.6)	RO4
	Experimental Evaluation • Conducted six experiment sessions • 190 participants • Two experimental settings - One group used the tool - The other without the tool (baseline)	• Group using the tool shows higher performance than the baseline. - (p-value = 0.03 for the statistical difference in Tool 1) - (p-value < 0.001 for the statistical difference in Tool 2)	
	Focus Group Evaluation • Conducted 2 focus groups • 10 participants - 6 crowdsourcing experts - 4 Ph.D. students with related backgrounds	• Strong evidence that the tool provides structured information • Mixed evidence that the tool frames and changes participants' decisions • Strong evidence of ease of use • A few suggestions for improvements of the tool	

7.2 Research Contributions

Having been a design science endeavour, our work contributes knowledge throughout its research activities, from problem definition, to sound research process, to solutions and their reflection, and to communication of the research results. Consequently, as it is very hard to provide an exhaustive list of all research contributions, we have identified four major contributions. Each of them is discussed in the following sections.

7.2.1 A New Approach for Establishing Crowdsourcing as an Organisational Business Process

At the beginning of the book, we noted that organisations face the challenge of how to establish crowdsourcing as an organisational business process. Despite a decade of research, most crowdsourcing research still relied heavily on an ad hoc perspective, studying individual aspects of the crowdsourcing process. In many cases, these studies explored and investigated crowdsourcing as a one-off process, rather than a common organisational practice. Consequently, the challenge still remains.

Our first approach to this challenge is *the introduction of a business process lens on crowdsourcing processes, designating the concept of BPC*. While the term BPC was coined in 2010 (La Vecchia & Cisternino, 2010), it was not widely used in the domain. It is this book that clarifies the BPC concept by balancing between the business process construct and the crowdsourcing construct. With BPC as a template, multiple instances of the same crowdsourcing process may be created. Our conceptualisation of BPC is partly theoretical, based on crowdsourcing literature and business process literature, and partly empirical, based on our observation that existing crowdsourcing processes have several activities that are repeatedly performed, as confirmed below.

The BPC conceptualisation can only stand if there are common repeatable activities of crowdsourcing processes. In this book, the condition has been satisfied. The book, through the scoping review, has confirmed that there is a set of common activities of the crowdsourcing processes, repeatedly found in multiple knowledge sources. These common activities, which have also been reinforced by other recent reviews (Amrollahi, 2015; Hosseini et al., 2015a), support the condition founding the BPC concept. Further, they suggest the main building blocks of BPC (Table 3.2).

Using the building blocks suggested by the scoping review, we *conceptualise BPC through a process model*. The model, on the one hand, clarifies the BPC conceptualisation through a process viewpoint with multiple structured activities that are necessary to establish crowdsourcing as an organisational business process. On the other hand, the model keeps the BPC conceptualisation focus. That is, the model focuses on the core repeatable building blocks of BPC, which defines the abstract structure of BPC. The abstract structure allows to build new crowdsourcing processes as real-life instances of the same core building blocks (Fig. 4.1). All in all, the process model, with its focus and business process lens, places BPC in a space quite distinct from one-off processes and their instances.

7.2.2 The Importance of the Ontology

Having introduced the concept of BPC, the book also proposes an ontology that offers knowledge structures around this concept. The ontology provides various unique benefits in BPC conceptualisation. We now discuss these benefits from three main research perspectives of the book: BPC, IS, and DSS.

Ontologies have played an important role in representing domains of knowledge (Fonseca & Martin, 2007; Guo, Schwartz, Burstein, & Linger, 2009; Wand & Weber, 1995). In this vein, our proposed *ontology represents the BPC domain*. More precisely, it defines BPC building blocks, processes, data, and data entities. It also structures the domain by presenting the hierarchical and decision-making relationships (Figs. 5.2 and 5.3). As a result, the ontology offers a scaffold for understanding the BPC domain. The representation of the ontology can be further characterised in two aspects: clarity (Akdemir et al., 2008; Fan et al., 2016) and coverage (Fan et al., 2016; Shanks et al., 2003).

The BPC ontology *has high clarity contributing to the understanding of the domain*, which can be seen via three points. First, it defines not only domain concepts but also hierarchical relationships and decision-making relationships, which increases shared understanding in the domain. Second, the ontology helps reduce semantic ambiguity. As noted previously, conflicting views and opinions exist in the domain, which leads to certain levels of semantic ambiguity. The ontology manages the conflicts through the 'wisdom of researchers', using the majority of knowledge sources as an indicator to address the conflicts. Finally, a combination of the ontology with the conceptual model and decision tool has provided three levels of abstraction for understanding the domain. All these points contribute to the high clarity of the ontology.

The BPC ontology also has *a high coverage of domain concepts and relationships*. This high coverage comes mainly from our grounded approach, which allows the ontological elements freely emerge. As a result, the ontological elements cover diverse aspects of the domain. We note however that in the grounding process, we made a decision that might reduce the coverage level of the ontology. That is, the decision to focus on the concepts supported by at least ten knowledge sources. Acknowledging the concern, we however have retained our decision since we have to balance the trade-off between coverage and complexity. Further, the evaluation of the ontology has lately shown that our decision is appropriate. More precisely, the comparison of our ontology with a version generated by OntoGen has shown that the BPC ontology broadly covers the domain. These results confirm the high coverage of the BPC ontology.

Before moving to the next perspective, we note here the *nature of our ontology*. If we follow Sharman et al. (2004) classifying ontologies as: top-level, domain, and application, our ontology should be seen as a domain ontology since we strictly focus on the BPC area. Furthermore, it should be treated as an informal ontology, rather than a formal one that would be defined using representation formalism languages. We nevertheless note that developing an informal ontology before transferring it into a formal one is a common, acceptable practice (Wong, Liu, & Bennamoun, 2012). Considering the BPC ontology in the lightweight-heavyweight continuum (Corcho et al., 2003), our work is aligned to the heavyweight ontologies since we examine not only concepts but also decision-making relationships and business rules in the BPC domain. As a result, we have contributed a heavyweight informal ontology to the BPC domain.

The IS discipline also highlights the role of ontologies. While agreeing with the ontology roles aforementioned in the BPC perspective, the IS discipline, in particular design science, suggests the contributions of ontologies for building knowledge bases (Miah, Gammack, & Kerr, 2007; Miah et al., 2014; Osterwalder & Pigneur, 2004; Ostrowski, Helfert, & Gama, 2014). In the book, *the ontology has offered a BPC knowledge base.* It builds the knowledge base through structuring the key concepts, hierarchical structures, and decision-making relationships, from which knowledge can be inferred. Furthermore, the knowledge base role of the ontology has been clearly revealed when the ontology formed the basis for tool construction. This is because founding artefact construction is a distinctive characteristic of knowledge bases (Hevner & Chatterjee, 2010). We note that the knowledge base offered by the ontology should not be limited only to construct the decision tool, but can also be used to constructing other IS artefacts, e.g. artefacts to standardise crowdsourcing processes. In short, we offer an ontological knowledge base for IS artefact development in the BPC domain.

Finally, we consider the ontology from the DSS (decision support system) perspective. In DSS literature, we identify two main roles of ontologies. The first role views ontologies as vocabulary frameworks defining terms, concepts and decision alternatives for certain DSS environments (Chen, Chen, Hsu, & Li, 2011; Van Valkenhoef, Tervonen, Zwinkels, De Brock, & Hillege, 2013). The second role, extending the first one, views ontologies as reasoning means, which structure logics of the DSS solutions (Amailef & Lu, 2013; Gennari et al., 2003; Miah et al., 2007). The BPC ontology in the current study is aligned with the second role, *ontology-supported reasoning*, for three reasons. First, the ontology helps develop reasoning knowledge, which has been showed via the exemplar of the decision tables (Tables 6.1, 6.2, and 6.3). Second, the reasoning role is aligned with the knowledge base role of the ontology, mentioned earlier in the design science perspective. Lastly, the ontology was actually integrated into the decision tool as a reasoning module (Fig. 6.1), which confirms its role as the ontology-supported reasoning means.

Overall, the BPC ontology plays several critical roles in the current research. It is a domain ontology clarifying and covering the BPC domain. It also serves as a knowledge base consolidating the existing knowledge for IS artefact development. Furthermore, it as a heavyweight ontology supports reasoning, which has been operationalised in the decision tool. These roles suggest the value of the BPC ontology.

7.2.3   Empirical Findings

In the above discussion, we have discussed BPC conceptualisation and its ontological structure for establishing crowdsourcing processes. In addition to these theoretical efforts, the book also brings empirical results that provide evidence on how our theoretical work can be operationalised to improve the establishment of BPC. Empirically, we constructed a decision tool and assessed it using experiments and focus groups.

The successful construction of the tool means four things. First, the tool construction has proved that BPC can actually be operationalised in practice. Second, the tool, which was developed based on the BPC ontology, has demonstrated the feasibility of the ontology. That is, the ontology can be implemented in a working system. Third, the construction has created an instantiation artefact (Hevner et al., 2004; March & Smith, 1995), which is a decision tool providing a means for decision makers to establish BPC step-by-step and to guide them in this establishment. Finally, the tool enables concrete assessments of its utility towards BPC establishment.

A mixed method was used for empirical assessment of the tool. We deployed a sequence of (1) the experiments to test whether the tool is useful for improving performance on BPC establishment and (2) the focus group to understand what aspects of the tool's usefulness are perceived by the participants. In the experiments consisting of 190 participants, the findings confirm that the use of the tool leads to better performance on both functions of the tool: the decision to crowdsource and crowdsourcing process design. From the results, we suggest that the tool is useful for BPC establishment. We note that although both functions are useful, and both are statistically supported, the support for process design (p-value < 0.001) is stronger than for the decision to crowdsource (p-value = 0.03). The experimental results alone cannot explain the difference, which has addressed in the focus group evaluation. In summary, the experiments provide empirical evidence suggesting the usefulness of the tool. While this usefulness is supported statistically, some of its aspects should be further evaluated and discussed.

Serving our intention to further evaluate the tool, two focus groups were conducted to gain insights on what aspects of the tool utility were perceived by the participants. The focus group results show that the tool benefits in terms of structuring BPC establishment and providing additional information for making informed decisions. It is also found that participants when using the tool have a positive perception towards ease of use, and they suggest a few possible improvements. There are mixed results on whether the tool may change the participants' decisions. Overall, the focus group results are positive towards the tool utility. They also help as a support to compare with the experimental results, as presented below.

Together, the two evaluation results enable us to confirm the tool utility, using both quantitative, individual-based, and controlled experiments, as well as qualitative, group-based, and likely naturalistic focus groups. It is also interesting to discuss their complementary findings. The focus group findings suggest that the tool is more useful for providing additional information than for changing participants' decisions. This provides a possible explanation for the different levels of support for the tool's utility in the experimental results regarding the decision to crowdsource and process design. Possibly, the equal support regarding the decision to crowdsource comes from the moderate ability of the tool that might or might not change participants' decisions, while the strong support regarding the process design comes from the strength of the tool that provides additional structured information in the design process.

Overall, the importance of the book relies not only on theoretical efforts, but also on having as much empirical evidence as possible. We have discussed the evidence from the experiments with 190 participants, and two focus groups. Apart from these, other empirical evidence was also collected and incorporated into the research results, including case studies of two crowdsourcing projects, and a pilot experiment with 46 participants. As a result, the empirical results have complemented and supported our theoretical efforts on BPC establishment.

7.2.4 Progression of Business Process Crowdsourcing

So far, we have presented our theoretical and empirical contributions to BPC establishment, which are expected to move the development of the BPC concept forward. To clarify this movement, we examine the progression of the concept in comparison with the literature review in Chap. 2. In that chapter, we reviewed three main research strands: the broad concept of crowdsourcing, crowdsourcing classifications, and crowdsourcing processes and the research foci of BPC. At that time, the three review strands covered quite broad aspects of crowdsourcing to form a foundation for our research. It is instructive if we re-examine these strands, focusing only on the concept of BPC.

Focusing on the BPC concept, we propose five phases of the concept progression. These phases are shown in Fig. 7.3. In the first phase, research conceptualised the overarching crowdsourcing concept by specifying its ideas and definitions (Estellés-Arolas & González-Ladrón-de-Guevara, 2012; Howe, 2006a), but did not mention BPC. The second phase started to classify different elements (Schenk & Guittard, 2011; Zhao & Zhu, 2014), in order to structure the crowdsourcing domain. At first, these structures were simple, just focusing on particular crowdsourcing elements. Also in this phase, a large number of studies researched ad hoc foci of crowdsourcing, which created a 'shopping list' of individual elements. Only in the third phase, the high-level building blocks of crowdsourcing processes became available. A few researchers were able to combine the individual elements forming an abstract crowdsourcing process and its building blocks (Amrollahi, 2015; Pedersen et al., 2013; Zogaj et al., 2014). Some of these building blocks are abstract and repeatable, which can be synthesised into BPC building blocks.

The fourth phase is the ongoing position of BPC. The target of this phase is to conceptualise and model the BPC concept leading to the proposition of reference models and ontologies. This phase is the focus of the book. The book conceptualised BPC using the building blocks synthesised from the scoping review (Chap. 3). We developed a process model (Chap. 4) and a heavyweight ontology guiding BPC (Chap. 5), which together provide a solid knowledge base for BPC establishment. Apart from our work, this phase also includes other recent models (Hetmank, 2014; Tranquillini et al., 2015), which enact and implement business processes based on crowdsourcing. Collectively, since this phase consists of our work that provides means to conceptualise, analyse, and design BPC, and the other work that provides means to enact and implement BPC (Hetmank, 2014; Tranquillini et al., 2015), this phase offers a solid scaffold supporting the whole

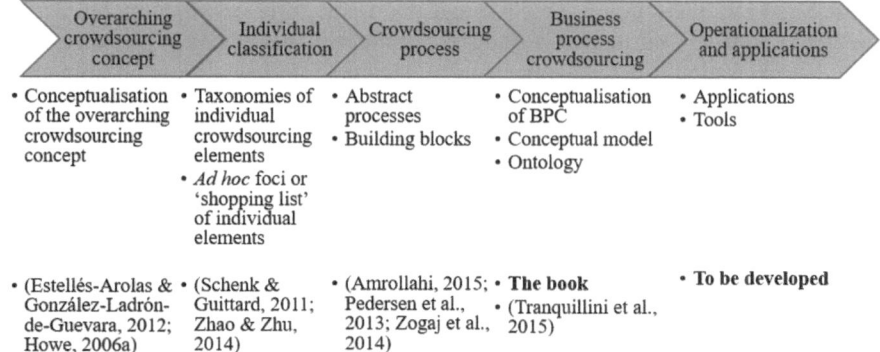

Overarching crowdsourcing concept	Individual classification	Crowdsourcing process	Business process crowdsourcing	Operationalization and applications
• Conceptualisation of the overarching crowdsourcing concept	• Taxonomies of individual crowdsourcing elements • *Ad hoc* foci or 'shopping list' of individual elements	• Abstract processes • Building blocks	• Conceptualisation of BPC • Conceptual model • Ontology	• Applications • Tools
• (Estellés-Arolas & González-Ladrón-de-Guevara, 2012; Howe, 2006a)	• (Schenk & Guittard, 2011; Zhao & Zhu, 2014)	• (Amrollahi, 2015; Pedersen et al., 2013; Zogaj et al., 2014)	• **The book** • (Tranquillini et al., 2015)	• **To be developed**

Fig. 7.3 Progression of business process crowdsourcing

business process based on crowdsourcing, from analysis, to design, and to implementation.

In the last phase, the models and ontologies of the previous phase can be applied to IS applications. Although our decision tool is an example of such applications (Chap. 6), we suggest that this is a to-be-developed area where diverse BPC tools and applications should be developed.

Overall, these five phases reflect the progression and expected development of BPC. They show the evolution of the domain, from an overarching concept, to individual structures, to abstract processes, to business process crowdsourcing, and to diverse BPC applications. Through this evolution of BPC, we think that the domain will continue progressing and further providing more applications to benefit organisations.

In summary, this section showed our contributions to the BPC domain. The contributions include the introduction of BPC conceptualisation, important roles of the ontology, empirical findings that show how our work operationalise and supports BPC, and progression of BPC. Together, they allow us to suggest that the book has contributed to move BPC forward in its progression in order to actually become an organisational business process.

7.3 Contributions to Practice

From a practical point of view, this book provides several practical contributions for organisations, decision makers, process designers, and project managers. The study provides organisations practical insights how to establish business processes based on crowdsourcing. In particular, organisations can use the conceptual model (Fig. 4.1) and ontology (Figs. 5.2 and 5.3) as a blueprint for analysing, planning and deploying crowdsourcing processes. The model provides defined steps on how to establish a crowdsourcing process; and the ontology presents structured activities, data, and data attributes in order to accomplish these steps. Together, they enable organisations to take advantage by integrating crowdsourcing into their organisational business processes.

Another practical contribution comes from the proposed decision framework (Fig. 4.2) and the set of decision tables (Tables 6.1, 6.2 and 6.3). They support decision makers to evaluate whether crowdsourcing is an appropriate strategy. The framework guides decision makers on what factors should be considered when making crowdsourcing decisions. Based on the framework, the decision tables formulate decision rules, which interpret and ease the decision-making process (Huysmans et al., 2011). As a result, we suggest that organisations can use the decision framework and decision tables as a practical means to measure their readiness for crowdsourcing.

The study provides a computer-based tool supporting BPC establishment. The tool structures concepts, relationships, business rules, and what-if scenarios, which supports managers and process designers in their BPC decision. This practical support is highlighted in particular through the experiments, where the results show that the tool can improve decision makers' performance in both the decision to crowdsource and process design. Furthermore, while the tool supports are mostly important to process designers, they may also be relevant to crowdsourcing platforms. By examining the tool, platform developers can integrate similar supports to assist their crowdsourcing customers.

Finally, one interesting implication for the use of the tool comes from the focus group results, which show that the tool can remind users of certain crowdsourcing aspects that they forgot. This implies that the tool can be used for cross checking crowdsourcing projects. In particular, the tool can advise project managers what aspects that should be focused and what are possibly missing in their projects. Managers can also compare their project plan with what have been suggested by the tool in order to analyse and monitor the projects. This use of cross checking is further highlighted as the tool has been launched as a web tool, ready for managers to visit and exercise their crowdsourcing projects.

7.4   Limitations of the Research

Through a critical lens, the study reported in this book inevitably still has certain limitations. First, we understand the risk of building a knowledge base from very diverse knowledge sources, whose bias and limitations may be transferred to the knowledge base (Kitchenham, 2007). Understanding this concern, we however note that the use of diverse knowledge sources benefits from the 'wisdom of researchers', which utilises diverse opinions for developing a more comprehensive view of particular phenomena like BPC (Surowiecki, 2004).

Another limitation comes from our decision to choose the cut-off value of ten knowledge sources when applying the 'wisdom of researchers'. This decision might exclude some interesting concepts and relationships in the 'long tail' that were supported by less than ten sources. We nevertheless note that this decision was made in order to balance between complexity and representation. If the chosen value was low, the complexity would increase since many concepts would be selected (Jonker & Pennink, 2010). In contrast, if the value was high, the representation would reduce since only a few building blocks would be selected. After testing different values, we finally chose ten as the cut-off value that balances complexity and representation.

There is another limitation related to the development of the decision tool, which focuses on "proof of concept" prototypes. The tool was developed through the rapid prototyping method, and thus targeted only at the level of evaluation and demonstration. Although the tool can be redesigned to meet industry targets, future research could implement the tool by applying proper software engineering methods. Besides, when we experimented with the tool, it was recognised that using students as proxies for crowdsourcing decision makers would be a limitation. Yet, we note that the use of students to experiment with software tools is an acceptable practice. Sjøberg et al. (2005) survey 113 software controlled experiments and show that "87 percent of the subjects were students" (p. 751). Furthermore, we have addressed the limitation with complementary data, where we used focus groups with crowdsourcing experts in order to triangulate our results.

7.5 Conclusion and Future Work

In this book we have presented our efforts towards establishing crowdsourcing as an organisational business process, particularly in establishing Business Process Crowdsourcing, BPC. Returning to our starting point, we have observed that organisations are often unsure about the way to best structure crowdsourcing activities and integrate them with other internal business processes. It also seems that this challenge comes from the predominant view in the domain that crowdsourcing is a one-off process. Furthermore, promising research stream from business, such as the use of a business process lens, has scarcely been adopted in the domain.

Addressing the challenge, one main contribution of the book is the introduction of BPC that views crowdsourcing as an integrated business process, rather than a one-off process. We have established BPC from the design-centric approach in that the majority of our work is centred on the iterations of design and evaluation. These iterations bring to the domain several IS artefacts, starting with a knowledge base constructed from scoping knowledge sources. Based on the knowledge base, we propose and validate a process model guiding organisations to manage the main building blocks of BPC establishment. Building on the process model, we propose an ontology that structures the BPC domain. It consists of the concepts, hierarchical relationships, and decision-making relationships necessary to establish crowdsourcing as an organisational business process. We note that both the process model and ontology are founded on the knowledge base. Thus, they represent a synthesis of the domain knowledge and as a result add a step forward to the conceptual and ontological structure of the BPC domain.

As a benefit of the ontological approach, it enables us to implement a tool that assists managers and process designers addressing the complexity of BPC establishment. The tool helps make informed decisions in BPC establishment, including decisions in adopting, designing and configuring novel crowdsourcing business processes. Regarding its evaluation, the tool was assessed through experiments and focus groups, which have shown positive results towards the utility. These results, together with other evaluations throughout the research, suggest that the decision tool together with the conceptual model and BPC ontology should be utilised to establish BPC.

Overall, *our conclusions from this book are positive towards the establishment of crowdsourcing as an organisational business process.* The conceptual model, ontology, and decision tool, constructed and validated in the research, should be used to support the establishment. While some of these artefacts have been presented in our independent publications (Thuan et al., 2016, 2017, 2018; Thuan et al., 2015), it is this book that structure them into a set of integrated knowledge, which has a strong theoretical ontological basis and promising empirical results. Consequently, we offer a body of knowledge for business process crowdsourcing, as a first attempt to establish the chosen phenomenon. By doing so, we hope that our attempt will motivate future researchers to investigate this important BPC domain. In this hope, we outline below a number of possible paths for future research.

Future Work
This book opens several paths for further exploring the potential of BPC and analogous to the general research field of crowdsourcing. Future research should use the ontological elements: the concepts, hierarchical relationships, and decision-making relationships *to design crowdsourcing experiments and field studies.* In other words, the ontology serves as a basic for developing a broad research agenda in the area. In this agenda, additional research should focus on the decision-making relationships, given the low number of supporting sources for this type of relationships in the domain.

Future research should aim to move the knowledge provided by the artefacts built in this study forward to a higher level of abstraction with *BPC design theories.* This research direction, aligning with Gregor and Hevner (2013), suggests that with the proposed artefacts and instantiations, related design theories can be developed. Such BPC design theories can add generality to our proposed artefacts. For instance, the BPC process model has been grounded from knowledge sources in the domain, and thus it is expected to be applicable in a variety of BPC contexts. Therefore, future research should further apply the model in different contexts, which will show its application principles and thus provide a basis for a theory explaining and predicting its use.

Our work also presents large research *opportunities for further design-based efforts* in both academia and industry. In particular, as a solid knowledge base in the domain, the BPC ontology can be used to construct different artefacts. Some that we can think of at this point in time are knowledge-based and collaborative-based systems, which are some common applications based on ontologies (Chandrasekaran, Josephson, & Benjamins, 1999). Others, that only time could uncover, may emerge from the combination of interoperability, reasoning and knowledge organisation provided by the ontology.

From a technical perspective, while we have already proposed a decision tool for BPC, our work mainly focuses on the business process aspects of the crowdsourcing process. Thus, *it is interesting to further develop and integrate our work from a more technical standpoint.* We note that several toolkits that configure and program crowdsourcing processes have existed (Kittur et al., 2011; Pavel Kucherbaev et al., 2013; Little et al., 2010; Tranquillini et al., 2015). Given that, future research may investigate how to connect the decision-support focus in our work and existing technical toolkits. This connection would offer a decision support system that would assist organisations, from the time they analyse, model, and design BPC, until the time they instantiate it using a particular set of programming toolkits.

In conclusion of this book, it is clear that crowdsourcing has been an important sourcing strategy for organisations in the last decade, and this trend is expected to continue with business process crowdsourcing. By establishing crowdsourcing as an organisational business process, organisations can take full advantage of the strategy. This book proposes a set of BPC artefacts that supports the establishment. Furthermore, a solid knowledge base of BPC is built and enriched through theoretical, ontological, and empirical scaffoldings. This solid knowledge base is promising for the future development of the domain to progress towards a mature crowdsourcing strategy.

References

Adepetu, A., Ahmed, K. A., Al Abd, Y., Al Zaabi, A., & Svetinovic, D. (2012). CrowdREquire: A requirements engineering crowdsourcing platform. *2012 AAAI Spring Symposium Series.*

Afuah, A., & Tucci, C. L. (2012). Crowdsourcing as a solution to distant search. *Academy of Management Review, 37*(3), 355–375.

Afuah, A., Tucci, C. L., & Viscusi, G. (2018). *Creating and capturing value through crowdsourcing.* Oxford University Press.

Ågerfalk, P. J., & Fitzgerald, B. (2008). Outsourcing to an unknown workforce: Exploring opensurcing as a global sourcing strategy. *MIS Quarterly,* 385–409.

Ågerfalk, P. J., Fitzgerald, B., & Stol, K. J. (2015). *Crowdsourcing software sourcing in the age of open* (pp. 45–60). Springer International Publishing.

Akdemir, U., Turaga, P., & Chellappa, R. (2008). An ontology based approach for activity recognition from video. *Proceedings of the 16th ACM international conference on Multimedia* (pp. 709–712).

Allahbakhsh, M., Benatallah, B., Ignjatovic, A., Motahari-Nezhad, H. R., Bertino, E., & Dustdar, S. (2013). Quality control in crowdsourcing systems: Issues and directions. *IEEE Internet Computing, 17*(2), 76–81.

Allahbakhsh, M., Ignjatovic, A., Benatallah, B., Beheshti, S., Bertino, E., & Foo, N. (2012, 14–17 Oct. 2012). *Reputation management in crowdsourcing systems.* Paper presented at the 8th International Conference on Collaborative Computing: Networking, Applications and Worksharing (CollaborateCom).

Alonso, O. (2013). Implementing crowdsourcing-based relevance experimentation: An industrial perspective. *Information Retrieval, 16*(2), 101–120.

Alonso, O., & Baeza-Yates, R. (2011). Design and implementation of relevance assessments using crowdsourcing. In P. Clough, C. Foley, C. Gurrin, G. F. Jones, W. Kraaij, H. Lee, & V. Mudoch (Eds.), *LNCS* (Vol. 6611, pp. 153–164). Berlin, Heidelberg: Springer.

Amailef, K., & Lu, J. (2013). Ontology-supported case-based reasoning approach for intelligent m-Government emergency response services. *Decision Support Systems, 55*(1), 79–97.

Amrollahi, A. (2015). A process model for crowdsourcing: Insights from the literature on implementation. *Proceedings of the Australasian Conference on Information Systems 2015, Paper 18.*

Anderson, D. R., Sweeney, D. J., & Williams, T. A. (2011). *Statistics for business and economics.* South-Westerb, Cengage Learning.

Archak, N. (2010). Money, glory and cheap talk: Analyzing strategic behavior of contestants in simultaneous crowdsourcing contests on TopCoder.com. *Proceedings of the 19th International Conference on World Wide Web* (pp. 21–30).

Baba, Y., & Kashima, H. (2013). Statistical quality estimation for general crowdsourcing tasks. *Proceedings of the 19th ACM SIGKDD International Conference on Knowledge Discovery and Data Mining* (pp. 554–562).

© Springer International Publishing AG, part of Springer Nature 2019

N. H. Thuan, *Business Process Crowdsourcing*, Progress in IS,

https://doi.org/10.1007/978-3-319-91391-9

Bacon, C. J., & Fitzgerald, B. (2001). A systemic framework for the field of information systems. *ACM SIGMIS Database, 32*(2), 46–67.

Bailey, K. D. (1994). *Typologies and taxonomies: An introduction to classification techniques.* Sage.

Baskerville, R. L., & Myers, M. D. (2002). Information systems as a reference discipline. *MIS Quarterly,* 1–14.

Berente, N., Vandenbosch, B., & Aubert, B. (2009). Information flows and business process integration. *Business Process Management Journal, 15*(1), 119–141.

Boudreau, K. J., & Lakhani, K. R. (2013). Using the crowd as an innovation partner. *Harvard business review, 91*(4), 60–69.

Bozzon, A., Brambilla, M., Ceri, S., & Mauri, A. (2013). Reactive crowdsourcing. *Proceedings of the 22nd International Conference on World Wide Web* (pp. 153–164).

Brabham, D. C. (2008a). Crowdsourcing as a model for problem solving: An introduction and cases. *Convergence: The International Journal of Research into New Media Technologies, 14* (1), 75–90.

Brabham, D. C. (2008b). Moving the crowd at iStockphoto: The composition of the crowd and motivations for participation in a crowdsourcing application. *First monday, 13*(6).

Brabham, D. C. (2009). Crowdsourcing the public participation process for planning projects. *Planning Theory, 8*(3), 242–262.

Brabham, D. C. (2010). Moving the crowd at threadless. *Information, Communication & Society, 13*(8), 1122–1145.

Brabham, D. C. (2011). The myth of amateur crowds. *Information, Communication & Society, 15* (3), 394–410.

Brabham, D. C. (2012). Motivations for participation in a crowdsourcing application to improve public engagement in transit planning. *Journal of Applied Communication Research, 40*(3), 307–328.

Brabham, D. C. (2013). *Crowdsourcing.* Canbridge, MA: The MIT Press.

Brooks, J., McCluskey, S., Turley, E., & King, N. (2015). The utility of template analysis in qualitative psychology research. *Qualitative Research in Psychology, 12*(2), 202–222.

Buecheler, T., Sieg, J. H., Füchslin, R. M., & Pfeifer, R. (2010). Crowdsourcing, open innovation and collective intelligence in the scientific method: a research agenda and operational framework. *Proceedings of the twelfth international conference on the synthesis and simulation of living systems, Odense, Denmark.*

Burger-Helmchen, T., & Pénin, J. (2010). *The limits of crowdsourcing inventive activities: What do transaction cost theory and the evolutionary theories of the firm teach us.* Strasbourg, France: Paper presented at the Workshop on Open Source Innovation.

Carlsson, S. A., Henningsson, S., Hrastinski, S., & Keller, C. (2011). Socio-technical IS design science research: Developing design theory for IS integration management. *Information Systems and e-Business Management, 9*(1), 109–131.

Chanal, V., & Caron-Fasan, M. L. (2010). The difficulties involved in developing business models open to innovation communities: The case of a crowdsourcing platform. *M@n@gement, 13*(4), 318–340.

Chandler, D., & Kapelner, A. (2013). Breaking monotony with meaning: Motivation in crowdsourcing markets. *Journal of Economic Behavior & Organization, 90,* 123–133.

Chandrasekaran, B., Josephson, J. R., & Benjamins, V. R. (1999). What are ontologies, and why do we need them? *IEEE Intelligent Systems and their Applications, 14*(1), 20–26.

Chen, C. C., Chen, K., Hsu, C.-Y., & Li, Y.-C. J. (2011). Developing guideline-based decision support systems using protégé and jess. *Computer Methods and Programs in Biomedicine, 102* (3), 288–294.

Chesbrough, H. (2013). *Open business models: How to thrive in the new innovation landscape*: Harvard Business Press.

Chilton, L. B., Horton, J. J., Miller, R. C., & Azenkot, S. (2010). Task search in a human computation market. *Proceedings of the ACM SIGKDD Workshop on Human Computation* (pp. 1–9).

Chilton, L. B., Little, G., Edge, D., Weld, D. S., & Landay, J. A. (2013). Cascade: Crowdsourcing taxonomy creation. *Proceedings of the SIGCHI Conference on Human Factors in Computing Systems* (pp. 1999–2008).

Cooper, R. G. (2008). Perspective: The Stage-Gate® idea-to-launch process—Update, what's new, and NexGen systems*. *Journal of Product Innovation Management, 25*(3), 213–232.

Corcho, O., López, M. F., & Gómez-Pérez, A. (2003). Methodologies, tools and languages for building ontologies. Where is their meeting point? *Data & Knowledge Engineering, 46*(1), 41–64.

Corney, J., Torres-Sanchez, C., Jagadeesan, A. P., Yan, X., Regli, W., & Medellin, H. (2010). Putting the crowd to work in a knowledge-based factory. *Advanced Engineering Informatics, 24*(3), 243–250.

Cross, N. (1982). Designerly ways of knowing. *Design Studies, 3*(4), 221–227.

Dai, P., Lin, C. H., & Weld, D. S. (2013). POMDP-based control of workflows for crowdsourcing. *Artificial Intelligence, 202,* 52–85.

Delir Haghighi, P., Burstein, F., Zaslavsky, A., & Arbon, P. (2013). Development and evaluation of ontology for intelligent decision support in medical emergency management for mass gatherings. *Decision Support Systems, 54*(2), 1192–1204.

Dennis, A. R., & Valacich, J. S. (2001). Conducting experimental research in information systems. *Communications of the Association for Information Systems, 7*(1).

Dennis, A. R., & Valacich, J. S. (2014). A replication manifesto. *AIS Transactions on Replication Research, 1*(1), 1–4.

Denyer, D., & Tranfield, D. (2006). Using qualitative research synthesis to build an actionable knowledge base. *Management Decision, 44*(2), 213–227.

Dibbern, J., Goles, T., Hirschheim, R., & Jayatilaka, B. (2004). Information systems outsourcing: a survey and analysis of the literature. *ACM SIGMIS Database, 35*(4), 6–102.

DiPalantino, D., & Vojnovic, M. (2009). Crowdsourcing and all-pay auctions. *Proceedings of the 10th ACM Conference on Electronic Commerce* (pp. 119–128).

Djelassi, S., & Decoopman, I. (2013). Customers' participation in product development through crowdsourcing: Issues and implications. *Industrial Marketing Management, 42*(5), 683–692.

Doan, A., Ramakrishnan, R., & Halevy, A. Y. (2011). Crowdsourcing systems on the world-wide web. *Communications of the ACM, 54*(4), 86–96.

Eickhoff, C., & De Vries, A. (2013). Increasing cheat robustness of crowdsourcing tasks. *Information Retrieval, 16*(2), 121–137.

Estellés-Arolas, E., & González-Ladrón-de-Guevara, F. (2012). Towards an integrated crowd-sourcing definition. *Journal of Information Science, 38*(2), 189–200.

Fan, S., Hua, Z., Storey, V. C., & Zhao, J. L. (2016). A process ontology based approach to easing semantic ambiguity in business process modeling. *Data & Knowledge Engineering, 102,* 57–77.

Feller, J., Finnegan, P., Hayes, J., & O'Reilly, P. (2012). 'Orchestrating'sustainable crowd-sourcing: A characterisation of solver brokerages. *The Journal of Strategic Information Systems, 21*(3), 216–232.

Fettke, P., & Loos, P. (2003). Classification of reference models: A methodology and its application. *Information Systems and e-Business Management, 1*(1), 35–53.

Flostrand, A. (2017). Finding the future: Crowdsourcing versus the Delphi technique. *Business Horizons, 60*(2), 229–236.

Fonseca, F., & Martin, J. (2007). Learning the differences between ontologies and conceptual schemas through ontology-driven information systems. *Journal of the Association for Information Systems, 8*(2), Article 2.

Fortuna, B., Grobelnik, M., & Mladenic, D. (2007). OntoGen: Semi-automatic ontology editor. In M. Smith & G. Salvendy (Eds.), *Human interface and the management of information. Interacting in information environments* (Vol. 4558, pp. 309–318). Springer.

Freyne, J., Coyle, L., Smyth, B., & Cunningham, P. (2010). Relative status of journal and conference publications in computer science. *Communications of the ACM, 53*(11), 124–132.

Geiger, D., & Schader, M. (2014). Personalized task recommendation in crowdsourcing information systems—Current state of the art. *Decision Support Systems, 65*, 3–16.

Geiger, D., Seedorf, S., Schulze, T., Nickerson, R. C., & Schader, M. (2011). Managing the crowd: Towards a taxonomy of crowdsourcing processes. *Proceedings of the Seventeenth Americas Conference on Information Systems*, Paper 430.

Gennari, J. H., Musen, M. A., Fergerson, R. W., Grosso, W. E., Crubézy, M., Eriksson, H., et al. (2003). The evolution of Protégé: An environment for knowledge-based systems development. *International Journal of Human-Computer Studies, 58*(1), 89–123.

Ghezzi, A., Gabelloni, D., Martini, A., & Natalicchio, A. (2017). Crowdsourcing: A review and suggestions for future research. *International Journal of Management Reviews*.

Giachetti, R. E. (2004). A framework to review the information integration of the enterprise. *International Journal of Production Research, 42*(6), 1147–1166.

Gibson, M., & Arnott, D. (2007). The use of focus groups in design science research. *ACIS 2007 Proceedings*, Paper 14.

Gomes, C., Schneider, D., Moraes, K., & de Souza, J. (2012). Crowdsourcing for music: Survey and taxonomy. *2012 IEEE International Conference on Systems, Man, and Cybernetics (SMC)* (pp. 832–839).

Gonnokami, K., Morishima, A., & Kitagawa, H. (2013). Condition-task-store: A declarative abstraction for microtask-based complex crowdsourcing. *Proceedings of the 1st VLDB Workshop on Databases and Crowdsourcing* (pp. 20–25).

Grace, K., Maher, M. L., Preece, J., Yeh, T., Stangle, A., & Boston, C. (2015). A process model for crowdsourcing design: A case study in citizen science. In J. S. Gero & S. Hanna (Eds.), *Design computing and cognition'14* (pp. 245–262). Springer.

Gregor, S. (2006). The nature of theory in information systems. *MIS Quarterly, 30*(3), 611–642.

Gregor, S., & Hevner, A. R. (2013). Positioning and presenting design science research for maximum impact. *MIS Quarterly, 37*(2), 337–355.

Gruber, T. R. (1993). A translation approach to portable ontology specifications. *Knowledge acquisition, 5*(2), 199–220.

Guarino, N., Oberle, D., & Staab, S. (2009). What is an ontology? In S. Staab & R. Studer (Eds.), *Handbook on ontologies* (pp. 1–17). Springer.

Guo, T., Schwartz, D. G., Burstein, F., & Linger, H. (2009). Codifying collaborative knowledge: Using Wikipedia as a basis for automated ontology learning. *Knowledge Management Research & Practice, 7*(3), 206–217.

Hasselbring, W. (2000). Information system integration. *Communications of the ACM, 43*(6), 32–38.

Haythornthwaite, C. (2009). Crowds and communities: Light and heavyweight models of peer production. *42nd Hawaii International Conference on System Sciences HICSS '09* (pp. 1–10).

Hetmank, L. (2013). *Components and functions of crowdsourcing systems—A systematic literature review*. Paper presented at the 11th International Conference on Wirtschaftsinformatik, Leipzig, Germany.

Hetmank, L. (2014). *Developing an ontology for enterprise crowdsourcing*. Paderborn: Paper presented at the Multikonferenz Wirtschaftsinformatik.

Hevner, A., & Chatterjee, S. (2010). *Design research in information systems: Theory and practice. Integrated series in information systems* (Vol. 22). Berlin, Heidelberg: Springer.

Hevner, A., March, S. T., Park, J., & Ram, S. (2004). Design science in information systems research. *MIS Quarterly, 28*(1), 75–105.

Hirth, M., Hoßfeld, T., & Tran-Gia, P. (2011). *Anatomy of a crowdsourcing platform-using the example of microworkers.com*. Paper presented at the Fifth International Conference on Innovative Mobile and Internet Services in Ubiquitous Computing (IMIS), Seoul.

Hirth, M., Hoßfeld, T., & Tran-Gia, P. (2012). Analyzing costs and accuracy of validation mechanisms for crowdsourcing platforms. *Mathematical and Computer Modelling, 57*(11–12), 2918–2932.

Holsapple, C. W. (2008). DSS architecture and types. In F. Burstein & C. W. Holsapple (Eds.), *Handbook on decision support systems 1* (pp. 163–189). Springer.

Hosack, B., Hall, D., Paradice, D., & Courtney, J. F. (2012). A look toward the future: Decision support systems research is alive and well. *Journal of the Association for Information Systems, 13*(5), 315–340.

Hossain, M. (2012). *Users' motivation to participate in online crowdsourcing platforms*. Paper presented at the 2012 International Conference on Innovation Management and Technology Research (ICIMTR).

Hossain, M., Kauranen, I., & Busi, M. (2015). Crowdsourcing: A comprehensive literature review. *Strategic Outsourcing: An International Journal, 8*(1), 2–22.

Hosseini, M., Phalp, K., Taylor, J., & Ali, R. (2014). The four pillars of crowdsourcing: A reference model. *IEEE Eighth International Conference on Research Challenges in Information Science (RCIS), 2014*, 1–12.

Hosseini, M., Phalp, K., Taylor, J., & Ali, R. (2015). On the configuration of crowdsourcing projects. *International Journal of Information System Modeling and Design (IJISMD), 6*(3), 27–45.

Hosseini, M., Shahri, A., Phalp, K., Taylor, J., & Ali, R. (2015). Crowdsourcing: A taxonomy and systematic mapping study. *Computer Science Review, 17*, 43–69.

Hoßfeld, T., Keimel, C., Hirth, M., Gardlo, B., Habigt, J., Diepold, K., et al. (2013). Best practices for QoE crowdtesting: QoE assessment with crowdsourcing. *IEEE Transactions on Multimedia, 16*(2), 541–558.

Höst, M., Regnell, B., & Wohlin, C. (2000). Using students as subjects—A comparative study of students and professionals in lead-time impact assessment. *Empirical Software Engineering, 5* (3), 201–214.

Howe, J. (2006a, June). Crowdsourcing: A definition. *Crowdsourcing: Tracking the rise of the amateur (Weblog)*. Retrieved from http://crowdsourcing.typepad.com/cs/2006/06/crowdsourcing_a.html.

Howe, J. (2006b). The rise of crowdsourcing. *Wired Magazine 2006, 14*, 1–4.

Howe, J. (2008). *Crowdsourcing: How the power of the crowd is driving the future of business*. Century.

Huberman, B. A., Romero, D. M., & Wu, F. (2009). Crowdsourcing, attention and productivity. *Journal of Information Science, 35*(6), 758–765.

Huston, L., & Sakkab, N. (2006). Connect and develop. *Harvard Business Review, 84*(3), 58–66.

Huysmans, J., Dejaeger, K., Mues, C., Vanthienen, J., & Baesens, B. (2011). An empirical evaluation of the comprehensibility of decision table, tree and rule based predictive models. *Decision Support Systems, 51*(1), 141–154.

Ipeirotis, P. G., Provost, F., & Wang, J. (2010). Quality management on amazon mechanical turk. *Proceedings of the ACM SIGKDD Workshop on Human Computation* (pp. 64–67).

Jabareen, Y. (2009). Building a conceptual framework: Philosophy, definitions, and procedure. *International Journal of Qualitative Methods, 8*(4), 49–62.

Jeppesen, L. B., & Lakhani, K. R. (2010). Marginality and problem-solving effectiveness in broadcast search. *Organization Science, 21*(5), 1016–1033.

Jonker, J., & Pennink, B. W. (2010). Conceptual models. In *The essence of research methodology* (pp. 43–63). Berlin, Heidelberg: Springer.

Kannangara, S. N., & Uguccioni, P. (2013). Risk management in crowdsourcing-based business ecosystems. *Technology Innovation Management Review, 3*(12), 32–38.

Karger, D. R., Oh, S., & Shah, D. (2013). Efficient crowdsourcing for multi-class labeling. *ACM SIGMETRICS Performance Evaluation Review, 41*(1), 81–92.

Katz, R., & Allen, T. J. (1982). Investigating the not invented here (NIH) syndrome: A look at the performance, tenure, and communication patterns of 50 R&D Project Groups. *R&D Management, 12*(1), 7–20.

Kaufmann, N., Schulze, T., & Veit, D. (2011). More than fun and money. Worker motivation in crowdsourcing—A study on mechanical turk. *Proceedings of the Seventeenth Americas Conference on Information Systems, Detroit, MI*, Paper 340.

Kazai, G. (2010). An exploration of the influence that task parameters have on the performance of crowds. *Proceedings of the CrowdConf 2010*.

Kazai, G., Kamps, J., & Milic-Frayling, N. (2011). Worker types and personality traits in crowdsourcing relevance labels. *Proceedings of the 20th ACM International Conference on Information and Knowledge Management* (pp. 1941–1944).

Kazman, R., & Chen, H. M. (2009). The metropolis model a new logic for development of crowdsourced systems. *Communications of the ACM, 52*(7), 76–84.

Khazankin, R., Psaier, H., Schall, D., & Dustdar, S. (2011). QoS-based task scheduling in crowdsourcing environments. In G. Kappel, Z. Maamar, & H. Motahari-Nezhad (Eds.), *Service-oriented computing* (Vol. 7084, pp. 297–311). Springer.

Khazankin, R., Satzger, B., & Dustdar, S. (2012a). Optimized execution of business processes on crowdsourcing platforms. *IEEE 8th International Conference on Collaborative Computing: Networking, Applications and Worksharing* (pp. 443–451).

Khazankin, R., Satzger, B., & Dustdar, S. (2012b). Predicting QoS in scheduled crowdsourcing. In J. Ralyté, X. Franch, S. Brinkkemper, & S. Wrycza (Eds.), *Advanced information systems engineering* (Vol. 7328, pp. 460–472). Berlin, Heidelberg: Springer.

King, N. (2004). Using templates in the thematic analysis of texts. In C. Cassell & G. Symon (Eds.), *Essential guide to qualitative methods in organizational research* (pp. 256–270). SAGE Publications.

King, N. (2012). Doing template analysis. *Qualitative organizational research: Core methods and current challenges*, 426–450.

Kitchenham, B. (2007). Guidelines for performing systematic literature reviews in software engineering. *Ver. 2.3 EBSE Technical Report*.

Kittur, A., Nickerson, J., Bernstein, M., Gerber, E., Shaw, A., Zimmerman, J., Lease, M., & Horton, J. (2013). The future of crowd work. *Proceedings of the 2013 Conference on Computer Supported Cooperative Work*.

Kittur, A., Smus, B., Khamkar, S., & Kraut, R. E. (2011). Crowdforge: Crowdsourcing complex work. *Proceedings of the 24th Annual ACM Symposium on User Interface Software and Technology* (pp. 43–52).

Kleemann, F., Voß, G. G., & Rieder, K. (2008). Un (der) paid Innovators: The commercial utilization of consumer work through crowdsourcing. *Science, Technology & Innovation Studies, 4*(1), 5–26.

Kohlborn, T. (2012). *Identification and evaluation of service bundles for governmental one-stop portals.* (Doctor of Philosophy), Queensland University of Technology.

Kohler, T. (2015). Crowdsourcing-based business models: How to create and capture value. *California Management Review, 57*(4), 63–84.

Kohler, T., & Nickel, M. (2017). Crowdsourcing business models that last. *Journal of Business Strategy, 38*(2), 25–32.

Kordon, F. (2002). An introduction to rapid system prototyping. *IEEE Transactions on Software Engineering, 28*(9), 817–821.

Krueger, R. A., & Casey, M. A. (2014). *Focus groups: A practical guide for applied research.* Sage publications.

Kucherbaev, P., Daniel, F., Tranquillini, S., & Marchese, M. (2016). Crowdsourcing processes: A survey of approaches and opportunities. *IEEE Internet Computing, 2*(2), 50–56.

Kucherbaev, P., Tranquillini, S., Daniel, F., Casati, F., Marchese, M., Brambilla, M., et al. (2013). Business processes for the crowd computer. In M. L. Rosa & P. Soffer (Eds.), *Business process management workshops* (pp. 256–267). Berlin, Heidelberg: Springer.

Küçük, D., & Arslan, Y. (2014). Semi-automatic construction of a domain ontology for wind energy using Wikipedia articles. *Renewable Energy, 62,* 484–489.

Kulkarni, A., Can, M., & Hartmann, B. (2012). Collaboratively crowdsourcing workflows with turkomatic. *Proceedings of the ACM 2012 Conference on Computer Supported Cooperative Work* (pp. 1003–1012).

La Vecchia, G., & Cisternino, A. (2010). Collaborative workforce, business process crowdsourcing as an alternative of BPO. In F. Daniel & F. M. Facca (Eds.), *ICWE2010. LNCS* (Vol. 6385, pp. 425–430). Springer.

LaToza, T. D., & Hoek, A. v. d. (2016). Crowdsourcing in software engineering: Models, motivations, and challenges. *IEEE Software, 33*(1), 74–80.

Le, J., Edmonds, A., Hester, V., & Biewald, L. (2010). *Ensuring quality in crowdsourced search relevance evaluation: The effects of training question distribution.* Paper presented at the SIGIR 2010 workshop on crowdsourcing for search evaluation.

Leimeister, J. M., Huber, M., Bretschneider, U., & Krcmar, H. (2009). Leveraging crowdsourcing: Activation-supporting components for IT-based ideas competition. *Journal of Management Information Systems, 26*(1), 197–224.

Levy, Y., & Ellis, T. J. (2006). A systems approach to conduct an effective literature review in support of information systems research. *Informing Science: International Journal of an Emerging Transdiscipline, 9,* 181–212.

Lim, Y.-K., Stolterman, E., & Tenenberg, J. (2008). The anatomy of prototypes: Prototypes as filters, prototypes as manifestations of design ideas. *ACM Transactions on Computer-Human Interaction (TOCHI), 15*(2), Article 7.

Little, G., Chilton, L. B., Goldman, M., & Miller, R. C. (2010). Turkit: Human computation algorithms on mechanical turk. *Proceedings of the 23nd Annual ACM Symposium on User Interface Software and Technology* (pp. 57–66).

Lloret, E., Plaza, L., & Aker, A. (2012). Analyzing the capabilities of crowdsourcing services for text summarization. *Language Resources and Evaluation, 47*(2), 337–369.

Lofi, C., Selke, J., & Balke, W. T. (2012). Information extraction meets crowdsourcing: A promising couple. *Datenbank-Spektrum, 12*(2), 109–120.

Lopez, M., Vukovic, M., & Laredo, J. (2010). PeopleCloud service for enterprise crowdsourcing. *IEEE International Conference on Services Computing (SCC), 2010,* 538–545.

López, M. F., Gómez-Pérez, A., & Corcho, O. (2004). *Ontological engineering: With examples from the areas of knowledge management, e-commerce and the semantic web.* Springer.

Lu, B., Hirschheim, R., & Schwarz, A. (2015). Examining the antecedent factors of online microsourcing. *Information Systems Frontiers, 17*(3), 601–617.

Lüttgens, D., Pollok, P., Antons, D., & Piller, F. (2014). Wisdom of the crowd and capabilities of a few: Internal success factors of crowdsourcing for innovation. *Journal of Business Economics, 84*(3), 339–374.

Majchrzak, A., & Malhotra, A. (2013). Towards an information systems perspective and research agenda on crowdsourcing for innovation. *The Journal of Strategic Information Systems, 22*(4), 257–268.

Malone, T. W., Laubacher, R., & Dellarocas, C. (2010). The collective intelligence genome. *IEEE Engineering Management Review, 38*(3), 38–52.

Man-Ching, Y., King, I., & Kwong-Sak, L. (2011). A survey of crowdsourcing systems. *2011 IEEE Third International Conference on Privacy, Security, Risk and Trust (passat), and 2011 IEEE Third International Conference on Social Computing (socialcom)* (pp. 766–773).

Mao, K., Capra, L., Harman, M., & Jia, Y. (2017). A survey of the use of crowdsourcing in software engineering. *Journal of Systems and Software, 126,* 57–84.

March, S. T., & Smith, G. F. (1995). Design and natural science research on information technology. *Decision Support Systems, 15*(4), 251–266.

Marjanovic, S., Fry, C., & Chataway, J. (2012). Crowdsourcing based business models: In search of evidence for innovation 2.0. *Science and Public Policy, 39*(3), 318–332.

Marsh, H. W. (2007). Students' evaluations of university teaching: Dimensionality, reliability, validity, potential biases and usefulness. In R. P. Perry & J. C. Smart (Eds.), *The scholarship of teaching and learning in higher education: An evidence-based perspective* (pp. 319–383). Springer.

Mason, W., & Suri, S. (2012). Conducting behavioral research on Amazon's mechanical turk. *Behavior research methods, 44*(1), 1–23.

Mason, W., & Watts, D. J. (2009). Financial incentives and the "performance of crowds". *Proceedings of the ACM SIGKDD Workshop on Human Computation* (pp. 77–85).

Mettler, T., Eurich, M., & Winter, R. (2014). On the use of experiments in design science research: A proposition of an evaluation framework. *Communications of the Association for Information Systems, 34*(1), 223–240.

Miah, S. J. (2008). *an ontology based design environment for rural business decision support.* (Doctor of Philosophy), Griffith University, Brisbane, Australia.

Miah, S. J., Gammack, J., & Kerr, D. (2007). Ontology development for context-sensitive decision support. *Third International Conference on Semantics, Knowledge and Grid* (pp. 475–478).

Miah, S. J., Kerr, D., & von Hellens, L. (2014). A collective artefact design of decision support systems: Design science research perspective. *Information Technology & People, 27*(3), 259–279.

Miah, S. J., Kerr, D. V., & Gammack, J. G. (2009). A methodology to allow rural extension professionals to build target-specific expert systems for Australian rural business operators. *Expert Systems with Applications, 36*(1), 735–744.

Miles, M. B., Huberman, A. M., & Saldaña, J. (2014). *Qualitative data analysis: A methods sourcebook.* SAGE Publications, Incorporated.

Mingers, J. (2003). The paucity of multimethod research: A review of the information systems literature. *Information Systems Journal, 13*(3), 233–249.

Montgomery, C. D. (2012). *Design and analysis of experiments* (8th ed.). New York: John Willey & Sons Inc.

Muhdi, L., Daiber, M., Friesike, S., & Boutellier, R. (2011). The crowdsourcing process: An intermediary mediated idea generation approach in the early phase of innovation. *International Journal of Entrepreneurship and Innovation Management, 14*(4), 315–332.

Muntés-Mulero, V., Paladini, P., Manzoor, J., Gritti, A., Larriba-Pey, J. L., & Mijnhardt, F. (2013). Crowdsourcing for industrial problems. In J. Nin & D. Villatoro (Eds.), *Citizen in sensor networks. LNCS* (Vol. 7685, pp. 6–18). Berlin, Heidelberg: Springer.

Naderi, B. (2018). *Motivation of workers on microtask crowdsourcing platforms.* Springer.

Nakatsu, R. T., Grossman, E. B., & Iacovou, C. L. (2014). A taxonomy of crowdsourcing based on task complexity. *Journal of Information Science, 40*(6), 823–834.

Naroditskiy, V., Jennings, N. R., Van Hentenryck, P., & Cebrian, M. (2013). Crowdsourcing dilemma. *arXiv preprint arXiv:1304.3548.*

Nickerson, R. C., Varshney, U., & Muntermann, J. (2012). A method for taxonomy development and its application in information systems. *European Journal of Information Systems, 22*(3), 336–359.

Nielsen, J. (2006, October). Participation inequality: Encouraging more users to contribute. *Jakob Nielsen's alertbox.* Retrieved from https://www.nngroup.com/articles/participation-inequality/.

Okoli, C. (2015). A guide to conducting a standalone systematic literature review. *Communications of the Association for Information Systems, 37*(1), Article 43.

Okoli, C., & Schabram, K. (2010). A guide to conducting a systematic literature review of information systems research. *Sprouts: Working Papers on Information Systems, 10*(26).

OReilly, T. (2007). What is Web 2.0: Design patterns and business models for the next generation of software. *Communications & strategies,* (1), 17.

Osterwalder, A. (2004). *The business model ontology: A proposition in a design science approach.* (Doctor of Philosophy), Institut d'Informatique et Organisation. Lausanne, Switzerland, University of Lausanne, Ecole des Hautes Etudes Commerciales HEC.

Osterwalder, A., & Pigneur, Y. (2004). An ontology for e-business models. *Value creation from e-business models* (pp. 65–97).

Ostrowski, L., Helfert, M., & Gama, N. (2014). Ontology engineering step in design science research methodology: A technique to gather and reuse knowledge. *Behaviour & Information Technology, 33*(5), 443–451.

Palacios, M., Martinez-Corral, A., Nisar, A., & Grijalvo, M. (2016). Crowdsourcing and organizational forms: Emerging trends and research implications. *Journal of Business Research, 69*(5), 1834–1839.

Paolacci, G., Chandler, J., & Ipeirotis, P. (2010). Running experiments on amazon mechanical turk. *Judgment and Decision Making, 5*(5), 411–419.

Paré, G., Trudel, M.-C., Jaana, M., & Kitsiou, S. (2015). Synthesizing information systems knowledge: A typology of literature reviews. *Information & Management, 52,* 183–199.

Park, S., Shoemark, P., & Morency, L.-P. (2014). Toward crowdsourcing micro-level behavior annotations: The challenges of interface, training, and generalization. *Proceedings of the 19th International Conference on Intelligent User Interfaces* (pp. 37–46).

Pedersen, J., Kocsis, D., Tripathi, A., Tarrell, A., Weerakoon, A., Tahmasbi, N., Jie, X., Wei, D., Onook, O., & De Vreede, G. J. (2013). Conceptual foundations of crowdsourcing: A review of IS research. *46th Hawaii International Conference on System Sciences (HICSS)* (pp. 579–588).

Peffers, K., Rothenberger, M., Tuunanen, T., & Vaezi, R. (2012). Design science research evaluation. In K. Peffers, M. Rothenberger, & B. Kuechler (Eds.), *Design science research in information systems. Advances in theory and practice. LNCS* (Vol. 7286, pp. 398–410). Berlin, Heidelberg: Springer.

Pfeiffer, J., Benbasat, I., & Rothlauf, F. (2014). Minimally restrictive decision support systems. *ICIS 2014 Proceedings*, Paper 36.

Poetz, M. K., & Schreier, M. (2012). The value of crowdsourcing: Can users really compete with professionals in generating new product ideas? *Journal of Product Innovation Management, 29* (2), 245–256.

Ranade, G., & Varshney, L. R. (2012). *To crowdsource or not to crowdsource?* Paper presented at the AAAI Workshop Human Comput. (HCOMP'12).

Rockwell, J. A., Grosse, I. R., Krishnamurty, S., & Wileden, J. C. (2010). A semantic information model for capturing and communicating design decisions. *Journal of Computing and Information Science in Engineering, 10*(3), 1–8.

Rosen, P. A. (2011). Crowdsourcing lessons for organizations. *Journal of Decision Systems, 20*(3), 309–324.

Rouse, A. C. (2010). A preliminary taxonomy of crowdsourcing. *Proceedings of the 21st Australasian Conference on Information Systems*, Paper 76.

Sakamoto, Y., Tanaka, Y., Yu, L., & Nickerson, J. V. (2011). The crowdsourcing design space. In D. Schmorrow & C. Fidopiastis (Eds.), *Foundations of augmented cognition. Directing the future of adaptive systems. LNCS* (Vol. 6780, pp. 346–355). Berlin, Heidelberg: Springer.

Sánchez, D., & Moreno, A. (2008). Learning non-taxonomic relationships from web documents for domain ontology construction. *Data & Knowledge Engineering, 64*(3), 600–623.

Sandkuhl, K., Smirnov, A., & Ponomarev, A. (2016). *Crowdsourcing in business process outsourcing: An exploratory study on factors influencing decision making.* Paper presented at the International Conference on Business Informatics Research.

Satzger, B., Psaier, H., Schall, D., & Dustdar, S. (2011). Stimulating skill evolution in market-based crowdsourcing. In S. Rinderle-Ma, F. Toumani, & K. Wolf (Eds.), *BPM 2011. LNCS* (Vol. 6896, pp. 66–82). Berlin, Heidelberg: Springer.

Satzger, B., Psaier, H., Schall, D., & Dustdar, S. (2012). Auction-based crowdsourcing supporting skill management. *Information Systems, 38*(4), 547–560.

Saxton, G. D., Oh, O., & Kishore, R. (2013). Rules of crowdsourcing: Models, issues, and systems of control. *Information Systems Management, 30*(1), 2–20.

Schall, D. (2012). *Service-oriented crowdsourcing: Architecture, protocols and algorithms.* Springer Science & Business Media.

Schenk, E., & Guittard, C. (2009). *Crowdsourcing: What can be outsourced to the crowd, and why?* Paper presented at the Workshop on Open Source Innovation, Strasbourg, France.

Schenk, E., & Guittard, C. (2011). Towards a characterization of crowdsourcing practices. *Journal of Innovation Economics, 7*(1), 93–107.

Schenk, E., Guittard, C., & Pénin, J. (2017). Open or proprietary? Choosing the right crowdsourcing platform for innovation. *Technological Forecasting and Social Change.*

Schulze, T., Seedorf, S., Geiger, D., Kaufmann, N., & Schader, M. (2011). Exploring task properties in crowdsourcing-an empirical study on mechanical turk. *ECIS2011, Paper 122.*

Seltzer, E., & Mahmoudi, D. (2013). Citizen participation, open innovation, and crowdsourcing challenges and opportunities for planning. *Journal of Planning Literature, 28*(1), 3–18.

Shanks, G., Tansley, E., & Weber, R. (2003). Using ontology to validate conceptual models. *Communications of the ACM, 46*(10), 85–89.

Sharman, R., Kishore, R., & Ramesh, R. (2004). Computational ontologies and information systems II: Formal specification. *Communications of the Association for Information Systems, 14*(1), Article 9.

Simula, H. (2013). The rise and fall of crowdsourcing? *46th Hawaii International Conference on System Sciences (HICSS)* (pp. 2783–2791).

Sjøberg, D. I., Hannay, J. E., Hansen, O., Kampenes, V. B., Karahasanovic, A., Liborg, N.-K., et al. (2005). A survey of controlled experiments in software engineering. *IEEE Transactions on Software Engineering, 31*(9), 733–753.

Smith, M., Busi, M., Ball, P., & Van Der Meer, R. (2008). Factors influencing an organisation's ability to manage innovation: A structured literature review and conceptual model. *International Journal of Innovation Management, 12*(04), 655–676.

Soh, C., Markus, M. L., & Goh, K. H. (2006). Electronic marketplaces and price transparency: Strategy, information technology, and success. *MIS Quarterly*, 705–723.

Sonnenberg, C., & vom Brocke, J. (2012a). Evaluation patterns for design science research artefacts. In M. Helfert & B. Donnellan (Eds.), *Practical aspects of design science* (Vol. 286, pp. 71–83). Berlin, Heidelberg: Springer.

Sonnenberg, C., & vom Brocke, J. (2012b). Evaluations in the science of the artificial–reconsidering the build-evaluate pattern in design science research. In K. Peffers, M. Rothenberger, & B. Kuechler (Eds.), *Design science research in information systems. Advances in theory and practice. LNCS* (pp. 381–397). Springer.

Stewart, O., Lubensky, D., & Huerta, J. M. (2010). Crowdsourcing participation inequality: A SCOUT model for the enterprise domain. *Proceedings of the ACM SIGKDD Workshop on Human Computation* (pp. 30–33).

Stol, K. J., & Fitzgerald, B. (2014). Two's company, three's a crowd: A case study of crowdsourcing software development. *Proceedings of the 36th International Conference on Software Engineering* (pp. 187–198).

Stol, K. J., LaToza, T. D., & Bird, C. (2017). Crowdsourcing for software engineering. *IEEE Software, 34*(2), 30–36.

Surowiecki, J. (2004). *The wisdom of crowds: Why the many are smarter than the few and how collective wisdom shapes business.* New York: Doubleday.

Sutherlin, G. (2013). A voice in the crowd: Broader implications for crowdsourcing translation during crisis. *Journal of Information Science, 39*(3), 397–409.

Tetreault, J., Chodorow, M., & Madnani, N. (2014). Bucking the trend: Improved evaluation and annotation practices for ESL error detection systems. *Language Resources and Evaluation, 48* (1), 5–31.

Threadless. (2015). Threadless help center. Retrieved from http://support.threadless.com.

Thuan, N. H., Antunes, P., & Johnstone, D. (2013). Factors influencing the decision to crowdsource. In P. Antunes, M. Gerosa, A. Sylvester, J. Vassileva, & G. J. De Vreede (Eds.), *CRIWG 2013. LNCS* (Vol. 8224, pp. 110–125). Heidelberg: Springer.

Thuan, N. H., Antunes, P., & Johnstone, D. (2014). Toward a nexus model supporting the establishment of business process crowdsourcing. In T. K. Dang, R. Wagner, E. Neuhold, M. Takizawa, J. Küng, & N. Thoai (Eds.), *FDSE 2014. LNCS* (Vol. 8860, pp. 136–150). Heidelberg: Springer.

Thuan, N. H., Antunes, P., & Johnstone, D. (2016). Factors influencing the decision to crowdsource: A systematic literature review. *Information Systems Frontiers, 18*(1), 47–68.

Thuan, N. H., Antunes, P., & Johnstone, D. (2017). A process model for establishing business process crowdsourcing. *Australasian Journal of Information Systems, 21,* 1–21.

Thuan, N. H., Antunes, P., & Johnstone, D. (2018). A decision tool for business process crowdsourcing: Ontology, design, and evaluation. *Group Decision and Negotiation, 27*(2), 285–312.

Thuan, N. H., Antunes, P., Johnstone, D., & Ha, X. S. (2015). Building an enterprise ontology of business process crowdsourcing: A design science approach. *PACIS 2015 Proceedings. AISeL, Paper 112.*

Tokarchuk, O., Cuel, R., & Zamarian, M. (2012). Analyzing crowd labor and designing incentives for humans in the loop. *IEEE Internet Computing Magazine, 16*(5), 45–51.

Tomlinson, B., Ross, J., Paul, A., Eric, B., Donald, P., Joseph, C., Martin, M., Syavash, N., Marco, L., Birgit, P., Andrew, T., David, C., Gary, O., Six, S., Marcus, S., Fabio, R. P., Albert, A. S., Eric, M., Xavier, F., Florian, F. M., Joseph, J. K., Rebecca, W. B., Marisa, L. C., Patrick, C. S., Johanna, B., Nitesh, G., Pirjo, N., Jeff, H., Nilufar, B., & Craig, S. (2012). Massively distributed authorship of academic papers. *Proceedings of the 2012 ACM annual conference extended abstracts on Human Factors in Computing Systems Extended Abstracts.*

Tranquillini, S., Daniel, F., Kucherbaev, P., & Casati, F. (2015). Modeling, enacting, and integrating custom crowdsourcing processes. *ACM Transactions on the Web (TWEB), 9*(2), Article 7.

Tremblay, M. C., Hevner, A. R., & Berndt, D. J. (2010). Focus groups for artifact refinement and evaluation in design research. *Communications of the Association for Information Systems, 26,* Article 27.

Tremblay, M. C., Hevner, A. R., & Berndt, D. J. (2012). Design of an information volatility measure for health care decision making. *Decision Support Systems, 52*(2), 331–341.

Uschold, M., & King, M. (1995). *Towards a methodology for building ontologies.* Paper presented at the IJCAI95 Workshop on Basic Ontological Issues in Knowledge Sharing, Montreal.

Valaski, J., Malucelli, A., & Reinehr, S. (2012). Ontologies application in organizational learning: A literature review. *Expert Systems with Applications, 39*(8), 7555–7561.

Van Aken, J. E. (2005). Management research as a design science: Articulating the research products of mode 2 knowledge production in management. *British Journal of Management, 16* (1), 19–36.

Van Aken, J. E., & Romme, A. G. L. (2012). A design science approach to evidence-based management. In D. M. Rousseau (Ed.), *The Oxford handbook of evidence-based management* (pp. 43–60).

van der Aalst, W., & Hee, K. M. (2004). *Workflow management: Models, methods, and systems.* Cambridge, MA: The MIT Press.

Van Valkenhoef, G., Tervonen, T., Zwinkels, T., De Brock, B., & Hillege, H. (2013). ADDIS: A decision support system for evidence-based medicine. *Decision Support Systems, 55*(2), 459–475.

Venable, J., Pries-Heje, J., & Baskerville, R. (2012). A comprehensive framework for evaluation in design science research. *Design Science Research in Information Systems. Advances in Theory and Practice,* 423–438.

Venable, J., Pries-Heje, J., & Baskerville, R. (2016). FEDS: A framework for evaluation in design science research. *European Journal of Information Systems, 25*(1), 77–89.

Venkatesh, V., & Davis, F. D. (2000). A theoretical extension of the technology acceptance model: Four longitudinal field studies. *Management Science, 46*(2), 186–204.

Vicente, K. J. (1999). *Cognitive work analysis: Toward safe, productive, and healthy computer-based work*: CRC Press.

Vogrinčič, S., & Bosnić, Z. (2011). Ontology-based multi-label classification of economic articles. *Computer Science and Information Systems, 8*(1), 101–119.

Vukovic, M. (2009). *Crowdsourcing for enterprises.* Paper presented at the 2009 World Conference on Services-I, Los Angeles, CA.

Vukovic, M., Laredo, J., & Rajagopal, S. (2010). Challenges and experiences in deploying enterprise crowdsourcing service. In B. Benatallah, F. Casati, G. Kappel, & G. Rossi (Eds.), *ICWE 2010. LNCS* (Vol. 6189, pp. 460–467). Berlin, Heidelberg: Springer.

Vuurens, J. B., & De Vries, A. P. (2012). Obtaining high-quality relevance judgments using crowdsourcing. *IEEE Internet Computing, 16*(5), 20–27.

Wand, Y., & Weber, R. (1995). On the deep structure of information systems. *Information Systems Journal, 5*(3), 203–223.

Wang, A., Hoang, C. D. V., & Kan, M.-Y. (2013). Perspectives on crowdsourcing annotations for natural language processing. *Language resources and evaluation, 47*(1), 9–31.

Webster, J., & Watson, R. T. (2002). Analyzing the past to prepare for the future: writing a literature review. *MIS Quarterly, 26*(2), xiii–xxiii.

Westpac. (2013, October). Westpac New Zealand to crowdsource mobile banking apps. Retrieved from http://www.westpac.co.nz/who-we-are/newsroom/media-releases-2013/westpac-new-zealand-to-crowdsource-mobile-banking-apps/.

Wexler, M. N. (2011). Reconfiguring the sociology of the crowd: Exploring crowdsourcing. *International Journal of Sociology and Social Policy, 31*(1/2), 6–20.

Whitla, P. (2009). Crowdsourcing and its application in marketing activities. *Contemporary Management Research, 5*(1), 15–28.

Wohlin, C., Runeson, P., Höst, M., Ohlsson, M. C., Regnell, B., & Wesslén, A. (2012). *Experimentation in software engineering.* Springer Science & Business Media.

Wong, W., Liu, W., & Bennamoun, M. (2012). Ontology learning from text: A look back and into the future. *ACM Computing Surveys (CSUR), 44*(4), Article 20.

Wu, F., Wilkinson, D. M., & Huberman, B. A. (2009). Feedback loops of attention in peer production. *International Conference on Computational Science and Engineering CSE'09* (pp. 409–415).

Wu, W., Tsai, W.-T., & Li, W. (2013). An evaluation framework for software crowdsourcing. *Frontiers of Computer Science, 7*(5), 694–709.

Wüllenweber, K., Beimborn, D., Weitzel, T., & König, W. (2008). The impact of process standardization on business process outsourcing success. *Information Systems Frontiers, 10*(2), 211–224.

Xu, Y., Ribeiro-Soriano, D. E., & Gonzalez-Garcia, J. (2015). Crowdsourcing, innovation and firm performance. *Management Decision, 53*(6), 1158–1169.

Yang, J., Adamic, L. A., & Ackerman, M. S. (2008). Crowdsourcing and knowledge sharing: Strategic user behavior on taskcn. *Proceedings of the 9th ACM Conference on Electronic commerce* (pp. 246–255).

Yin, R. K. (2013a). *Case study research: Design and methods.* Sage publications.

Yin, R. K. (2013). Validity and generalization in future case study evaluations. *Evaluation, 19*(3), 321–332.

Zachman, J. A. (1987). A framework for information systems architecture. *IBM systems journal, 26*(3), 276–292.

Zhao, Y., & Zhu, Q. (2014). Evaluation on crowdsourcing research: Current status and future direction. *Information Systems Frontiers, 16*(3), 417–434.

Zheng, H., Li, D., & Hou, W. (2011). Task design, motivation, and participation in crowdsourcing contests. *International Journal of Electronic Commerce, 15*(4), 57–88.

Zogaj, S., Bretschneider, U., & Leimeister, J. M. (2014). Managing crowdsourced software testing: A case study based insight on the challenges of a crowdsourcing intermediary. *Journal of Business Economics, 84*(3), 375–405.